How to Cool the Planet

Books by Jeff Goodell

The Cyberthief and the Samurai

*Sunnyvale: The Rise and Fall
of a Silicon Valley Family*

*Our Story: Seventy-Seven Hours That Tested
Our Friendship and Our Faith*

*Big Coal: The Dirty Secret Behind
America's Energy Future*

*How to Cool the Planet: Geoengineering
and the Audacious Quest to
Fix Earth's Climate*

For Michele

Nature is always better when left to itself—
but for what purpose?

—ANSEL ADAMS

How to Cool the Planet

GEOENGINEERING
AND THE AUDACIOUS QUEST
TO FIX EARTH'S CLIMATE

Jeff Goodell

Houghton Mifflin Harcourt
BOSTON / NEW YORK
2010

551.6
GOO

www.hmhbooks.com

Library of Congress Cataloging-in-Publication Data
Goodell, Jeff.
How to cool the planet : geoengineering and the audacious quest
to fix earth's climate / Jeff Goodell.
p.   cm.
Includes bibliographical references and index.
ISBN 978-0-618-99061-0
1. Climatic changes. 2. Climatic changes—Environmental
aspects. 3. Global warming. 4. Engineering geology.
5. Environmental engineering.
I. Title.
QC981.8.C5G666 2010
551.6—dc22   2009046565

Book design by Brian Moore

Printed in the United States of America

DOC 10 9 8 7 6 5 4 3 2 1

Contents

How to Cool
the Planet

ONE

....

The Prophet

I GREW UP in California, where human ingenuity is a force of nature. Computers, the Internet, Hollywood, blue jeans, the Beach Boys — they are all inventions of my home state. The economic and cultural power of these things is obvious. What's less obvious is how they transformed the place that gave birth to them. Until the early 1970s, my hometown of Silicon Valley was mostly orchards and Victorian ranch houses, with rows of cherry and apricot trees that marked the coming of spring with delicate white and pink blossoms. During the PC revolution, I watched those orchards fall to make room for glassy high-tech office buildings. The hillside where I saw the footprint of a mountain lion in the 1970s is now cluttered with houses. Silicon Valley is still a beautiful place, but the blossoms are mostly gone, the sky is hazy, and the beaches are crowded. This is happening everywhere, of course — it's the story of modern life. And there are many upsides to this transformation, including the fact that the ideas and technologies born in California have been a great boon to humanity. But you have to be pretty obtuse to grow up in a place like Silicon Valley and not be aware that progress sometimes comes at a price.

I left the Valley in my midtwenties and moved to New York City to begin a career as a journalist. My connection to the Valley served me well. I spent the next decade or so writing about the business and culture of my hometown for publications such as *Rolling Stone* and the *New York Times Magazine*. But my perspective changed after I became the father of three kids. The future of digital culture was suddenly much less interesting to me than the survival of the human race. I spent a lot of time with climate scientists while I was reporting my previous book, which was about the coal industry. It was a sobering experience. I think of myself as an optimistic person, but the deeper you probe into the climate crisis, the darker the story gets. It's hard not to read it as a parable about the dangers of living in a high-tech society. (No matter how hard they tried, a world of hunter-gatherers could not cook the planet.) And it's harder still not to wonder whether the smartest, most technologically sophisticated creatures that ever existed on earth will figure out a solution for this looming catastrophe. My friends in Silicon Valley are sure we can. They believe we are one big idea—Thin film solar! Cellulosic ethanol! High-altitude wind power!—away from solving this crisis. I used to think that, too.

In early 2006, a friend emailed me an essay by Paul Crutzen that was about to be published in an academic journal. Crutzen is a Dutch atmospheric chemist who won the Nobel Prize for his pioneering research on the ozone hole in the atmosphere. In his note, my friend—a successful entrepreneur in the solar power industry—wrote: "Read this. We are in deep trouble. We're going to geoengineer the damn planet now!"

I may have heard the word "geoengineer" once or twice before, but I knew next to nothing about it, other than the fact that it generally referred to people with outlandish ideas about how to counteract global warming. I had a vague memory of reading an article about a handful of scientists—I imagined them toiling in a lab buried deep in a mountain somewhere in New Mexico—who wanted to launch mirrors into space or dump iron into the ocean in a desperate attempt to cool the earth. The title of Crutzen's essay certainly amused me: "Albedo Enhancement by Stratospheric Sulfur Injections: A Contribution to Resolve a Policy Dilemma?" The phrase "albedo enhancement" sounded like a procedure a surgeon might perform on a lonely middle-aged man.

When I started to read, however, I was captivated. The basic facts were familiar: carbon dioxide (CO_2) levels in the earth's atmosphere are rising to concentrations not seen in twenty million years, with no end in sight. Meanwhile, the earth's climate is warming even faster than scientists had predicted just a few years ago. What was new in Crutzen's paper — new to me, anyway — was the view that some of this accelerated warming was driven not only by high levels of CO_2 but also by the progress we have made in the fight against smog and other traditional pollutants. The tiny particles that cause some kinds of air pollution act like mirrors in the sky, reflecting sunlight away from the earth, which cools the planet. As we eliminate pollution, the particles vanish, letting us all breathe easier — but also letting more sunlight in, which heats up the earth ever faster. As Crutzen pointed out, by trying to save kids from asthma, we were inadvertently making the climate crisis worse.

What to do? Clean air is obviously a good thing: air pollution kills people. The simplest solution would be to cut greenhouse gas emissions. If anyone should have been confident that we could take bold action to address this problem, it should have been Crutzen. After all, he was in part responsible for the fact that the leading nations of the world had come together in the late 1980s to confront another global threat, the ozone hole. In that case, once the risk of ozone damage was clear, action was swift: an international treaty, the Montreal Protocol, was negotiated and signed in 1987, banning ozone-depleting substances. It was an inspiring example of political leaders from around the world coming together to confront a grave threat in a rational and decisive way. But when it came to dealing with greenhouse gases, Crutzen was not so sanguine that a political solution could be found. He understood that the problem of reducing greenhouse gases is far deeper and more complex than eliminating chlorofluorocarbons from refrigerators and air conditioners, in part because greenhouse gas emissions are, in some ways, a proxy for economic health and prosperity. In fact, Crutzen called the notion that industrialized nations would join together and significantly reduce emissions "a pious wish."

Instead, Crutzen offered a radical proposal: rather than focusing entirely on cutting greenhouse gas emissions, maybe it was time to think about addressing the potentially catastrophic consequences of global warming in a different way. If the problem is too much heat, an obvious solution would be to find a way to reduce that heat. One method to do that would be to increase the earth's reflectivity in ways that would not cause asthma attacks and kill people. As Crutzen knew as well as anyone, about 30 percent of the energy from sunlight that hits the earth is immediately reflected back into space, while the other 70 percent is trapped here by CO_2 and other greenhouse gases, warming the planet. If we could reflect just 1 or 2 percent more sunlight away from the earth's surface, it would be like popping up an umbrella on the beach on a hot summer day. Crutzen called it albedo enhancement ("albedo" is just another word for reflectivity).

There are lots of ideas about how one might deflect sunlight away from the planet, from launching mirrors into space to painting roofs white. But as Crutzen pointed out in his paper, the simplest way to do it might be to add a relatively small number of sulfate particles—you can think of them as dust—to the upper atmosphere. The dust would remain in the stratosphere for only a year or so before raining out—so any serious geoengineering scheme would require continuous injection. But unlike pollution in the lower atmosphere, which is where the nasty stuff we breathe resides, pumping a modest amount of particles into the upper atmosphere would pose little danger to human health. The effect they might have on the chemistry of the stratosphere, especially the ozone layer that protects the earth from the sun's ultraviolet light, was, Crutzen admitted, unclear. However, his preliminary calculations suggested that the risks were low.

Would it work? On a scientific level, there is nothing complicated about it. Light colors reflect sunlight; dark colors absorb it. That's why asphalt is hot on your bare feet and white clothes are popular in the summer. The same basic idea holds true for the planet. Anything that reflects sunlight (ice, white roofs, certain kinds of clouds and air pollution) contributes to cooling; anything that absorbs sunlight (open water, evergreen forests in northern latitudes, asphalt parking lots) contributes to heating.

In his paper, Crutzen talked specifically about the cooling effect of volcanoes. For years, scientists have known that the sulfate particles that volcanoes spew into the air are remarkably effective at scattering sunlight. If the eruption is large enough, they can have a global impact on temperatures. One of the most recent examples is Mount Pinatubo, a volcano in the Philippines that erupted in 1991, lowering the earth's temperature by a degree or so for several years. A more extreme example of the phenomenon is the so-called nuclear winter—a theory that was much debated in the 1980s, suggesting that a nuclear war could inject enough soot and particles into the atmosphere to block out the sun and send temperatures plummeting.

Crutzen didn't say how we might go about mimicking volcanoes to offset global warming, except to suggest that there are lots of ways to inject particles into the stratosphere, including spraying them out of high-altitude aircraft, pushing them up a long hose tethered to a stratospheric balloon, or even shooting them up into the sky with artillery. As far as engineering challenges go, it wouldn't be too difficult. And even more important, it would be cheap. In Crutzen's estimation, we could engineer the earth's climate for less than 1 percent of the annual global military budget.

This all sounded interesting and provocative. It took me a while, however, to grasp just how mind-bending Crutzen's proposal really was. Here was one of the world's top atmospheric scientists suggesting that the climate crisis was so urgent and potentially catastrophic that the only way to save ourselves might be by filling the stratosphere with man-made pollution from artificial volcanoes. Had it really come to this?

In the media world—at least the part of the media world that takes science seriously—Crutzen's essay raised a ruckus. For one thing, the whole idea of changing the reflectivity of the planet as a way to offset global warming sounded downright wacky, even coming from a serious guy like Crutzen. As for injecting particles into the stratosphere—wasn't the goal to clear the air, not further pollute it? Geoengineering seemed like an idea ripped out of the pages of a sci-fi novel, conjuring up associations with Dr. Evil and crazy Cold War physicists and the hubris of the techno-elite. Perhaps worst of all, Crutzen's argument implied that the whole strategy of relying on an international agreement to cut greenhouse gas emissions was misguided—or at least grossly insufficient.

This was not a message the world was ready to hear. *An Inconvenient Truth*, Al Gore's documentary about global warming, had been released the same summer, waking millions of people up to the compelling scientific evidence behind the climate crisis. Progressive politicians around the world were beginning a major push to reduce emissions, trying, at least in public, to give the appearance that they were eager to fulfill their commitment to the Kyoto Protocol, the international agreement to cut greenhouse gas emissions signed in 1997. In Europe, the first market for greenhouse gas emissions trading was just taking off. Financial analysts predicted that the market would someday become the largest in the world, with hundreds of billions of dollars' worth of emissions credits being swapped every year, creating a powerful incentive for power companies to cut pollution and reap the rewards.

In this context, Crutzen was a turncoat, a man who dared to betray the growing movement to fight global warming just at the moment when it was gaining momentum. "This sounds to me like a miracle fix cooked up by Big Oil to keep the masses fat, dumb and happy," one blogger commented. "You keep driving and we'll get some smart scientists to air-condition the planet!"

But Crutzen's logic was not easy to dismiss. If there was one thing I had learned from the four years I'd spent researching and writing about coal, the dirtiest of fossil fuels, it was that the world was not going to stop burning black rocks anytime soon. Coal-fired power plants generate half the electricity in America. In the developing world, the percentage is even higher—India and China both get about 70 percent of their electricity from coal. The Chinese consume almost three times as much coal as we do in the United States—nearly three billion tons a year (although per capita, they consume far less).

Coal is the engine that is lifting people in the developing world out of poverty, not only giving them the power to light their homes and cook their food but also transforming them, for better or worse, into Western-style consumers. Unfortunately, coal is also the most carbon-intensive of fossil fuels, generating more than a third of the world's CO_2 pollution. Everyone wants to be hopeful about the possibilities of the renewable energy revolution, but the truth is, getting off coal in the near future—or, equally unlikely, figuring out a cheap and efficient way to burn coal without releasing CO_2 into the atmosphere—is a monumentally difficult challenge. And if we can't get off coal, it doesn't matter if every SUV driver rides a skateboard to work and Al Gore takes over as chairman of ExxonMobil—we won't have a hope in hell of staving off dangerous climate change.

Equally sobering were the encounters I had with climate scientists during my research. By 2006, the major scientific uncertainties about whether or not the planet was warming—and why it was warming—had long been settled. (I won't bother rehashing the evidence. If you still think global warming is a myth or unrelated to human activity, you're reading the wrong book.) But real questions remained, especially about the rate at which that warming would occur and what the impacts would be. A draft of the 2007 report by the Intergovernmental Panel on Climate Change (IPCC), the United Nations group comprising several hundred top scientists, was already circulating at the time Crutzen's essay appeared. The report predicted that the planet would warm by 3 to 7 degrees Fahrenheit by the end of the century and forecast a future of melting glaciers, rising seas, epic droughts, disease, and famine.

At the time, the report was touted as the first unequivocal statement from the scientific community about the cause and consequences of global warming. Off the record, however, many scientists were uncomfortable about how conservative the report was. There is good reason, of course, not to overstate scientific consensus or overhype impacts. But many felt it was equally dangerous — and even immoral, given the stakes — to underplay them.

Today it's clear that those scientists were right to be uncomfortable — the 2007 IPCC report is already woefully out-of-date. Global emissions are rising much faster than the report forecasted, and climate impacts are more severe. Credible studies now indicate that temperatures in the United States could increase by as much as 15 degrees Fahrenheit by the end of the century, with dust bowls in the Southwest and in many other heavily populated regions around the world. And instead of a sea level rise of less than a foot, as the IPCC report suggested, a number of respected climate modelers now believe it could be as high as three feet or more. James Hansen, the director of NASA's Goddard Institute for Space Studies and well-known as the godfather of global warming science, goes even further. He told me during a conversation in 2009 that if we don't cut emissions hard and fast, the seas could rise by as much as nine feet by the end of the century. Goodbye, Bangladesh, London, Miami — and Silicon Valley. If my grandchildren want to visit my hometown, they'll have to put on diving gear. "It would be a different planet," Hansen said.

You can see the increasing velocity of the changes most clearly in the Arctic, where winter temperatures in recent years have been as much as 3 degrees Fahrenheit warmer than average. Although that might not sound like a lot, a profound transformation of the region is already under way. According to the National Snow and Ice Data Center, the maximum extent of the summer sea ice cover for 2009 was the third-lowest on record. The six lowest maximum extents since satellite monitoring began in 1979 all occurred between 2004 and 2009. Extend this trend into the future, and the prognosis is not good. Back in 2006, many scientists were predicting that the Arctic would be ice-free in the summer by 2050. Now some scientists believe that it could happen within the next decade. "It's like man is taking the lid off the northern part of the planet," one polar ice expert commented. The loss of summer sea ice may well be a boon for shipping and oil exploration. But it is also likely to have broad impacts on the earth's climate, especially ocean circulation patterns, which, in turn, could disrupt major climate events such as the Asian and African monsoons. Nearly two billion people depend on those rains to grow their food.

One of the greatest misapprehensions about the climate crisis is the notion that we can fix all this simply by cutting emissions quickly. We can't. Even if we cut CO_2 pollution to zero tomorrow, the amount of CO_2 we have already pumped into the atmosphere will ensure that the climate will remain warm for centuries. To understand why, it's important to realize that CO_2 is not like the pollutants that create smog, most of which fall out of the air a few days or weeks after they are emitted. CO_2 lingers in the atmosphere for a very long time. Every time you drive to the store for a quart of milk, about 50 percent of the CO_2 you dump out of the tailpipe remains in the atmosphere for a decade or so before it is absorbed by the earth's carbon cycle. (The oceans are the single largest carbon-eaters, but plants and trees suck up a lot, too.) It takes a few centuries to absorb the next 30 percent. The final 20 percent lingers in the atmosphere for as long as 100,000 years.

The implications of this are profound. "The climatic impacts of releasing fossil fuel CO_2 to the atmosphere will last longer than Stonehenge," oceanographer David Archer wrote in 2008. "Longer than time capsules, longer than nuclear waste, far longer than the age of human civilization so far."

Besides sucking up carbon, the world's oceans play a big role in the response time of the climate, too. Their waters act like a giant heat sink for the planet. Once the oceans warm up, they will continue to radiate heat for hundreds of years. This thermal inertia, combined with CO_2's habit of hanging around in the atmosphere, means we may already be locked into dangerous levels of warming—we just don't know it yet.

Exactly how much does the climate warm with each additional ton of greenhouse gases we dump into the atmosphere? Scientists can make estimates, based on the heat-trapping properties of a CO_2 molecule, but the truth is, no one knows for sure. A key uncertainty in this calculation is the operation of feedback loops—that is, the mechanisms by which a dynamic system like the earth attempts to maintain equilibrium. Positive feedback loops are common. When ice melts in the Arctic, for example, it exposes more open water, which in turn absorbs more heat, which accelerates the warming, which melts more ice—you can see how these feedbacks build on each other. The question is, as positive feedbacks increase and the world heats up, what other changes might they trigger? One fear is that rapidly rising temperatures in the Arctic will cause the permafrost to thaw. Permafrost is loaded with methane, the byproduct of decomposing plants and microbes. If it melts quickly, it could send a sudden pulse of the gas into the atmosphere (methane is a short-lived but potent greenhouse gas, seventy times more powerful than CO_2), triggering a massive warming. Another fear is that changing rainfall patterns could cause tropical rain forests to collapse, eliminating a major carbon absorber, or carbon sink, for the planet. Many scientists have looked in vain for negative feedbacks—as yet undiscovered natural systems that will help to counteract the ever-increasing levels of greenhouse gases. So far, they haven't found any—at least not on a scale that matters.

Then there is the question of what Ken Caldeira, a climate modeler at the Carnegie Institution's Department of Global Ecology at Stanford University, calls the "ringiness" of the climate system. In other words, when you hit the system with a big hammer—and the thirty billion tons of CO_2 that humans dump into the atmosphere every year certainly qualify as a big hammer—how much noise does it make? Is this a system that can absorb a big hit, or does the impact amplify throughout the system, perhaps leading to unexpected gyrations? We know that in the past, the climate has jumped from one stable state to another. Consider a climatic event known as the Younger Dryas, which ended about 11,500 years ago, just as the earth was emerging from the last ice age. After several thousand years of warming, temperatures inexplicably plunged 10 to 15 degrees Fahrenheit and stayed that way for a thousand years, before warming up again just as suddenly. During the Younger Dryas, conditions in northern Europe were similar to those in the Arctic today. Icebergs drifted as far south as Portugal. Climate scientists aren't sure what caused this period of dramatic change, but what disturbs them is that if it happened once, it could happen again. Wally Broecker, a pioneering climatologist at Columbia University, has famously compared the earth's climate to a dragon: you poke it, and you're never sure how it is going to react. We could get lucky and be living in a climate system that is more tolerant than we think. But we could also be living in a system that is far more sensitive to being poked than we currently understand. By pushing the system so hard, we are, in effect, playing Russian roulette with the operating system of civilized life.

A few months after Crutzen's paper was published, I called him at his office at the Max Planck Institute for Chemistry in Mainz, Germany. We briefly discussed his concerns about the essay—a number of respected scientists had cautioned him not to publish it, fearing it would open a Pandora's box of questions about geoengineering—but Crutzen was determined to see it in print. "We live in a fool's climate," he told me. "We need to prepare for the worst. It might never occur. But if it does, and if we need to cool the earth off in a hurry, what will we do?"

At a dinner party not long ago, I found myself sitting next to the rare books librarian at a nearby college. She expressed concern, as many thoughtful people do, about the consequences of global warming. I mentioned to her that I had recently learned about a group of scientists who believe it might be time to deliberately engineer the earth's climate to counteract global warming. I talked about shooting particles into the sky, cloud machines, and mirrors in space. She looked at me for a moment, not sure whether I was serious or not, then burst out laughing. It was a reasonable response. You don't need a Ph.D. in physics to understand the basic insanity of this undertaking.

Some climate scientists have a slightly more nuanced view. Back in 2006, I discovered that talking about geoengineering with most scientists was like talking about mining minerals on Mars: an interesting idea, but not one they spent much time pursuing. It's not that they were morally against it. "Scientists love the idea of manipulating systems," Ken Caldeira told me. "It's how you learn how things work." The trouble, Caldeira explained, is that few scientists have the funding to simply explore projects that suit their fancies. And back in 2006, there were essentially zero dollars in the mainstream science world to study geoengineering. Why? You would think that Bush-era conservatives and climate skeptics—many of whom controlled the purse strings for science funding circa 2006—would have loved the idea. And a few did. In late 2001, just months after the Bush administration officially abandoned Kyoto and turned America into not only one of the world's biggest consumers of fossil fuels but also its most unabashed, the U.S. Department of Energy quietly convened a group of energy and climate experts to explore technological responses to rapid climate change. The report from the meeting—which included a pretty good rundown of the pluses and minuses of various geoengineering options—never saw the light of day. It's not hard to see why: to argue for geoengineering with any credibility, it must be combined with a clear call for reductions in greenhouse gas pollution. And until recently, very few conservatives have been willing to make that call.

Fear of ridicule was another barrier. Although the dream of manipulating the weather is almost as old as civilization itself, the idea of studying ways of deploying technology to manage the earth's climate was seen by some scientists as politically incorrect, dangerous, or just downright silly. Why put your career at risk when there are so many other more fruitful problems to explore? As a consequence, the researchers who explored geoengineering did so as a hobby or as a sideline to their main research projects. It was the scientific equivalent of a porn habit, something you thought about and explored in the privacy of your own lab but did not discuss in polite company, lest you be considered a pervert.

Indeed, many mainstream scientists were not shy about expressing the view that geoengineering was a crazy idea. "A semi-intelligent visitor from Mars would look down on us and say, You're all totally insane," said Vaclav Smil, a noted energy expert at the University of Manitoba in Canada. "The fact that we're even talking about it is a sign of desperation," Michael Oppenheimer, a climate scientist at Princeton, told me. "Foolishness," said Kevin Trenberth, the head of the Climate Analysis Section at the National Center for Atmospheric Research. John Holdren, who was head of Woods Hole Oceanographic Institution before he took his current job as chief science adviser to President Barack Obama, was equally dismissive. "The geoengineering approaches considered so far appear to be afflicted with some combination of high costs, low leverage, and a high likelihood of serious side effects," he wrote in 2006. And so it went, on and on. Beltway policy wonks and environmental leaders were even more emphatic in their opposition. David Hawkins, the head of the Climate Center at the Natural Resources Defense Council, called it "the Frankenplanet solution."

That was 2006. Today, in the aftermath of the 2009 Copenhagen climate summit, where world leaders talked about the dangers we face from global warming but failed to come up with a legally binding agreement to cut emissions, it's increasingly hard to cling to the idea that we're going to solve this problem with cooperation and good intentions. And as the rhetoric about cutting emissions grows increasingly hollow and the risks of rapid climate change grow ever higher, the public debate is shifting from how to stop global warming to how we can live with it. This means not only reengineering our drinking water infrastructure to prepare for earlier snowmelts and longer droughts and tracking new disease vectors in developing countries, but also beginning to think hard about what we might do if the worst-case scenarios came true and we really were faced with severe droughts that led to famines, or rapidly rising sea levels that threatened major cities and coastal regions. "If we had the climatic equivalent of the subprime economic meltdown, people would demand action," said Caldeira, who has done many key modeling studies on the impacts of various geoengineering schemes. "The temptation to throw some dust into the stratosphere to cool the planet off in a hurry might be hard to resist." In this context, geoengineering starts to look less like the fevered dream of mad scientists and more like the fevered dream of panicked politicians. And that is perhaps an even more frightening scenario.

Nonetheless, the old taboos are fading fast. In 2009, the British Royal Society, one of the most respected scientific organizations in the world, released a major geoengineering study. The U.S. equivalent of the Royal Society, the National Academy of Sciences, is likely to follow suit. Although it's tough to gauge how much support the idea has within President Obama's administration, Steven Chu, the secretary of energy, told me that "geoengineering is certainly worth further research." He often proselytizes for white roofs on buildings, which, besides saving energy on air conditioning, help (very slightly) to cool the planet by reflecting more sunlight than traditional dark roofs. One of Chu's top appointments at the Department of Energy, Steven Koonin, has long been intrigued by geoengineering and in fact chaired a weeklong scientific conference on the subject in 2008, when he was chief scientist for BP, the London-based petroleum giant. Even the U.S. Congress is sticking its toe in the water. In late 2009, the House Committee on Energy and

Technology held the first-ever hearings on geoengineering, signaling that the idea is now safe for public consumption. In addition, money for geoengineering research is starting to flow from private sources, including venture capitalists and technology-loving philanthropists such as Bill Gates. "I'm concerned about the impact of global warming on poor people in the developing world," Gates told me. "If geoengineering can help reduce that, I think it's worth exploring."

In the past year or two, climate skeptics and friends of the fossil fuel industry have also discovered geoengineering. The American Enterprise Institute, which has a long history of working to deny the scientific consensus on climate change and maintains strong ties to the fossil fuel industry (Lee Raymond, former CEO of ExxonMobil, served on its board; ExxonMobil was also a big financial supporter of its climate work), runs one of the few funded policy centers on geoengineering. Other organizations and personalities well-known for derailing serious action on global warming—such as the Cato Institute and Danish statistician Bjørn Lomborg, head of the Copenhagen Consensus Center—have pitched geoengineering as a cheap alternative to cutting emissions. As Alex Steffen, cofounder of Worldchanging, a popular environmental policy and activism website, wrote in April 2009, "Combining dire warnings about climate action's economic costs with exaggerated claims about geoengineering's potential is the new climate denialism."

In some ways, geoengineering is a lot like cloning, or genetic engineering, or even nanotechnology. It scrambles old political alliances and carves out new ideological fault lines. It stokes fears about Big Science run amuck, about the limits of human knowledge, about technological progress pushing beyond moral progress. But there are important differences, too. Unlike in these other cutting-edge scientific endeavors, nobody is actually doing any geoengineering yet. At least not in a deliberate way. You could certainly make the argument—and many people do—that civilization itself is a geoengineering project. (Crutzen famously dubbed the past ten thousand years or so "the Anthropocene.") But the difference is intention. For the past ten thousand years, we could be excused for behaving like locusts, unaware of the larger consequences of our all-consuming appetites. But that excuse is gone now—at least for all of us here in the land of iPhones and air conditioning. Our current greed and recklessness are starting to look a lot like a suicidal impulse.

At any science conference where geoengineering is discussed, it's a good bet that several hours will be devoted to a tortuous debate over what the term "geoengineering" actually means and whether the name should be changed to something softer and less technocratic, such as "climate intervention" or "climate restoration." For the purposes of this book, I'm going to stick with "geoengineering," simply because it's the current term of art. As for what exactly the word refers to, the most succinct definition I've encountered comes from that Royal Society paper: "the deliberate large-scale intervention in the Earth's climate system, in order to moderate global warming."

And before I go any deeper into this, I must tell you that a lot of the wacky ideas proposed by wannabe geoengineers will not be covered in this book. Two of my favorites: dumping millions of tons of Special K cereal into the ocean to change the reflectivity of the water and ionizing CO_2 molecules with lasers so that the earth's magnetic field ejects them from the atmosphere. Crazy ideas, of course, can be fun to read about. But I'm limiting the focus of this book to serious ideas that may be workable in the near term—by which I mean the next fifty years or so. That also eliminates some interesting but wildly impractical proposals, such as launching mirrors into space to deflect sunlight. We may indeed do something like that someday—building what amounts to a permanent louvered sunshade for the earth—but the cost and complexity of such a project are so huge that it is extremely unlikely that it would happen before the end of the century.

Finally, it's important to make clear from the outset that if we want to geoengineer the planet, there are two fundamentally different ways to do it. One method, sometimes called carbon engineering, focuses on CO_2 removal, which includes any technique or technology that pulls CO_2 out of the atmosphere, from dumping iron into the ocean in order to stimulate plankton blooms (which in turn absorb CO_2) to building scrubbers that remove CO_2 from the air. This is by far the least controversial of the two approaches to geoengineering, in part because it's slow acting and essentially mimics the earth's natural carbon cycle. (You could argue that large-scale tree plantations are a form of carbon engineering, too.)

The second method is to engineer the earth's albedo—to cool the planet by changing its reflectivity, as Crutzen suggested. On the simplest level, you could paint roofs and roads white. Or you could inject particles into the stratosphere, which would be much more effective. Another way would be to build a fleet of machines that would brighten clouds, causing them to scatter more sunlight. In contrast with carbon engineering, albedo engineering—or solar radiation management, as some scientists call it—could be useful in a climate emergency because it would cool the earth off instantly, in the same way that stepping into the shade on a hot day gives you quick relief from the sunlight. Interestingly, you don't have to deflect much sunlight to have a big impact. To offset a doubling of CO_2 levels from preindustrial conditions (a common benchmark among climate scientists), you would have to scatter just 2 percent of the light that hits the planet.

But messing around with the earth's albedo would also be far more ethically fraught than CO_2 removal. Deflecting sunlight is not a replacement for reducing CO_2 levels in the atmosphere. For one thing, shielding sunlight might lower the temperature, but it would do nothing to solve other problems related to high CO_2 levels, such as ocean acidification, which is killing coral reefs and could have a devastating impact on the ocean food chain. Also, because shielding sunlight would change the way heat is distributed around the earth and reduce the differential between daytime and nighttime temperatures, it would likely have broad climatic effects, such as shifting rainfall patterns and intensity. (Of course, unchecked global warming is also likely to have a big effect on rainfall.) Perhaps the most dangerous aspect of this kind of geoengineering is that throwing a few million tons of dust into the stratosphere would be relatively quick, cheap, and easy to do, putting it well within reach of petty dictators and tyrants.

If we ever got serious about geoengineering the planet, we would be likely to deploy a variety of both carbon engineering and sunlight-blocking technologies. The real question is not *how* we would do it. It's *should* we do it. I discovered that the case against deliberately manipulating the earth's climate really boils down to three arguments.

First, we are messing with a system we don't understand. The earth's climate is immeasurably complex—we may have fancy climate models and a passing understanding of atmospheric chemistry, but our ignorance remains vast. For example, scientists understand the basic physics of cloud formation, but beyond that, much of what happens in the clouds remains a mystery. If we don't even understand clouds, how can we hope to understand the complex interactions between earth and sky that shape the highly dynamic system we call climate? As David Battisti, a climate modeler at the University of Washington, put it, "Think of the climate like your car—the faster you go, the more likely you are to feel the wheels start to wobble. And the more you push it, the faster you go, the more likely it is to spin out of control." If the twentieth century taught us one thing, it's that technological innovations always have unintended consequences. Bomb builders at Los Alamos thought they were working on a device to bring world peace, but they also unleashed nuclear proliferation and fear of annihilation. The automobile brought personal freedom, but it also gave us strip malls, suburbs, and global warming. Dams control floods, create reservoirs, and generate clean power. They also destroy rivers, threaten migratory fish, and encourage mindless consumption of water.

What kind of havoc might geoengineering bring? You name it. Environmentally, the biggest concerns are shifting precipitation patterns and unexpected droughts, which could be devastating for many food-producing regions of the world. Foreign policy experts fear conflicts over nations "stealing" one another's rain. Military leaders worry about climate warfare. Human rights activists can already see a world in which the rich will use their technological superiority to screw the poor. If it comes down to a choice between rain in Africa or rain in Iowa, which region do you think is going to win?

Then there are the psychological consequences. What happens when the color of the sky on a particular day is the result not of Mother Nature's mood but of the mood of geoengineers who are spreading dust in the stratosphere? (Because of their light-scattering effects, high-altitude particles are likely to cause paler skies but more brightly colored sunsets.) What happens to our romance with Nature—sentimental as it may be at times—when we become hyperconscious that we are all living in a terrarium?

The second argument against geoengineering is that even talking about it distracts us from the urgent job of cutting greenhouse gas pollution. "It's like the way Baptists view sex education in school," Steve Rayner, professor of science and civilization at the University of Oxford, said at a recent geoengineering conference. "The worry is, if you start talking about it, you're more likely to do it." According to this view, if people believe there is a quick technological fix out there for global warming, they will ask why we should bother going through all the pain and struggle of reinventing the world's energy systems. After all, who wants to pay higher electric bills, move to a smaller house, or give up their third TV if we can just throw some dust in the air and cool off the planet? This is a version of the classic "moral hazard" argument that economists use frequently to underscore why flood insurance encourages people to build homes

in flood-prone locations, or why bank bailouts discourage investment firms from instituting real reforms. If someone else is going to
cover the loss, it greatly lessens the urgency of taking responsibility
for one's own actions. And what if the fix doesn't work as planned?
It's one thing to smoke in bed, confident that if anything happens,
you can dial 911 and the fire department will be there in two minutes. But what if when the firefighters arrive, the pumps and fire
hoses don't work?

In fact, betting on geoengineering as a substitute for cutting
emissions is a really bad idea for a number of reasons. For one, carbon engineering alone is unlikely to soak up more than a modest
percentage of the CO_2 we dump into the atmosphere burning fossil fuels. And when it comes to deflecting sunlight, there is what Ray
Pierrehumbert, a climate researcher at the University of Chicago,
calls "the Sword of Damocles problem." If we begin throwing dust
into the stratosphere to scatter sunlight, we have to keep doing it.
Unlike CO_2, particles that are injected into the stratosphere will fall
out of the sky after a year or so. If we don't inject more, the sky will
clear, which will trigger a sudden warming—just like stepping out
of the shadows into the sunlight. So it comes down to a question
of intent. If we inject particles into the sky as a way to keep Greenland from melting too quickly, or to buy us a few extra decades to
reduce emissions, it might make sense. If it's simply a lazy way of
continuing with business as usual, however, we're likely to make
our problems much worse. "If we keep emitting greenhouse gases
with the intent of offsetting the global warming with ever-increasing loadings of particles in the stratosphere, we will be heading to
a planet with extremely high greenhouse gases and a thick stratospheric haze that we would need to maintain more or less indefi-
nitely," Caldeira told me. "This seems to be a dystopian world out of
a science fiction story."

The third and, in my view, strongest argument against geoengineering is that it's evidence of exactly the same kind of industrial
thinking that cooked the planet in the first place. The fundamental challenge of global warming goes far beyond just cutting emis-

sions. It involves changing everything about our lives, from where we live and how we work to our definitions of progress and prosperity. And this is the real problem with geoengineering. Instead of reducing our voracious appetites for material goods and inspiring us to lead simpler lives, it compels us to chase after a technological fix — a high-tech Band-Aid that will solve all our problems.

The simple and obvious fact is that Western civilization as we know it is unsustainable. We are running out of cheap oil; we are overfishing the oceans; we are depleting our soils; we are running short of drinking water. Geoengineering is not going to fix all that. The only way to build a sustainable planet is to fundamentally reinvent our lives, either accelerating forward into a radically different future or falling back into a more primitive past. The virtue of riding out the climate crisis uncushioned by geoengineering, some would argue, is that it could provide the shock we need to sober up. That's an ugly way to think about it, but certainly there are many people who hold this view, whether they consciously admit it or not.

These are all compelling arguments for why geoengineering is a dangerous idea. But they aren't compelling enough for me to dismiss it entirely. Yes, progress is always a devil's bargain, but I am suspicious of full-frontal attacks on the evils of technological fixes. Catalytic converters, which remove pollutants from the tailpipes of cars, have saved tens of millions of lives. Antibiotics kill infections. Mosquito nets prevent malaria. Prosthetic knees, hips, and legs give people mobility. Milk from genetically engineered goats is used to manufacture lifesaving drugs. Computers process information — including enormously complex climate models, which allow us to speculate intelligently about just how much trouble our technological society has gotten itself into.

Similarly, geoengineering could turn out to be an important tool for risk reduction. If the climate crisis turns out to be even worse than we can now imagine, reflecting sunlight away from the planet would be one of the few options we would have to cool things off in a hurry. This doesn't mean we should start building cloud-brightening machines tomorrow, but it certainly does suggest that it might be prudent at least to begin to research these options in earnest — if only so that we can discover what *doesn't* work now before we urgently need a fix.

Still, the idea of pursuing geoengineering further — of taking a few years to write a book about it — did not really occur to me until I called David Keith, the head of the energy and environment program at the University of Calgary in Alberta, Canada. Keith, I knew, was a highly respected scientist studying ways to store CO_2 from coal-fired power plants underground in deep saline aquifers. What I didn't know was that he had also been thinking and writing about geoengineering for more than a decade. A quick telephone conversation convinced me that he knew as much about the moral, political, and engineering complexities of this idea as anyone I had encountered. More important, he was, at that very moment, building a machine to scrub CO_2 out of the atmosphere. "In fact, the machine will be up and running in a few days," he said. "You're welcome to come and see it if you want."

As a child of Silicon Valley, I've always had a soft spot for big thinkers and far-out hardware. And if you drew up a short list of inventions that could change the world, a machine that removes CO_2 would certainly be high on the list. It does not take much imagination to see that a machine like that — or, more accurately, a few thousand machines like that — could someday function as part of a climate control system. *How warm would you like your planet, sir?* Just set the thermostat at the CO_2 level you like, fine-tune the sunlight with some dust in the stratosphere or some machine-brightened clouds, and you have your Goldilocks climate — not too hot, not too cold, just right.

Of course, it wouldn't be quite that simple, but that was the promise. Was Keith's machine another step toward the cliff of extinction or a parachute for the human race? A week later, I jumped on a plane to Calgary to find out.

TWO

A Planetary Cooler

FLYING FROM NEW YORK to Calgary in September 2006, I spent a lot of time looking out my window at the passing landscape of the Midwest. From this vantage point, it's difficult to hold on to the fantasy that there is much "nature" left in the natural world. You see circles and squares of crops, a vast agricultural checkerboard where everything is leveled and contoured for maximum production. It's as distant from the old-fashioned idea of nature as the auto malls of California or Texas. These are vast monocultures of corn and wheat and sorghum, where even the inner workings of the crops have been tweaked by human ingenuity. The soil is just a platform, conceptually not so different from Windows 7 — an artificial environment built and maintained to support human needs. The crops themselves are as stimulated as any caffeine-addled cubicle jockey, fed a careful mixture of synthetic fertilizers that are the nutritional equivalents of PowerBars and protected by antivirus programs developed by companies such as Monsanto.

One way to view this landscape is as one of the highest achievements of human ingenuity and invention. If it were not for the transformation of these grains into a highly engineered commodity, there would be no hope of feeding the nearly seven billion people on the planet today. Environmentalists rightfully point out the devastating consequences of industrial farming, including soil erosion and fertilizer runoff that creates large toxic blooms in the Gulf of Mexico. But according to Nobel laureate Norman Borlaug, this kind of whining about the side effects of high-yield industrial farming is a luxury that only the rich can afford: "If [environmental lobbyists] lived just one month amid the misery of the developing world, as I have for fifty years, they'd be crying out for tractors and fertilizer and irrigation canals and be outraged that fashionable elitists back home were trying to deny them these things."

This view is violently contradicted by, among others, my wife, Michele, a passionate gardener who grows a good percentage of the vegetables our family eats. In her view, industrial farming is an unsustainable practice that has largely served the purposes of corporations, which have managed to achieve commodity control over our food supply. Michele looks at suburban lawns and sees wasted space where gardens should grow. She sees a world that is in danger of forgetting the most basic survival skill—how to put a seed in the ground and make it grow—and that depends on agricultural practices that claim to be modern and efficient but are in fact overengineered and planet deadening.

Somewhere over Wisconsin or Minnesota, it occurred to me that how you feel about geoengineering depends a lot on how you feel about what has happened to the world in the past two hundred years—whether you think the tools we have invented (and I mean that in the broadest sense, from toothbrushes and shovels to water treatment plants and nuclear weapons) have made the world a better place or not. We have made great progress in fighting disease and starvation. In the developing world, the child mortality rate has declined by more than a third in the past decade alone (thank you, Bill Gates!). In the West, we live longer, richer, more comfortable lives than any generation that preceded us. Far more people have access to far more knowledge (thank you, Google!) than at any other time in human history. But does all this mean we live *better* lives than our forebears? That is a philosophical question, and everyone has to answer it for himself or herself.

In places such as Silicon Valley, the reigning faith is all about pushing forward. Our problems today are not caused by technology, my friends in the Valley argue, but by big, dumb, old-fashioned technology, such as coal plants and V-8 engines. The future is not about cutting back but about moving forward faster and faster — ultraefficient electric cars, better photovoltaic cells, smarter grids, cheap water purification systems, genetically engineered crops, and all the other technological developments that will allow us to go on with our modern lives, but with radically reduced impacts. It's about scaling down our footprint and scaling up our consumer intelligence. It's about finding beauty in small things: iPod nanos, batteries built by viruses, carbon nanowires, homegrown tomatoes.

In this sense, geoengineering cuts against the spirit of the times. It's a big hammer for a big problem. It is about bringing human dominion to a new part of the planet. The best way to do this is not yet obvious, but what is obvious is that it can be done. We might use some version of David Keith's CO_2 scrubbers or genetically engineer crops to lighten the color of their leaves and reflect more sunlight back into space. Who knows what the future will bring? But there is nothing small about the idea, nothing modest, nothing of human scale. For better or worse, geoengineering is about turning the earth into the planetary equivalent of the industrialized farmland that was passing thirty thousand feet below me.

I had never been to Calgary before. I imagined the city would resemble Cheyenne, Wyoming, an old cow town on the high plains that feels a lot like yesterday. I was right about the high plains part — but that's it. The big city that sprawls out beneath the Canadian Rockies has the smell of a modern boomtown — tall office buildings rising downtown, sprawling housing developments on the outskirts, lots of expensive cars and SUVs crawling in heavy traffic. What's driving the bustling economy here is not software development or manufacturing — it's oil. Calgary is headquarters for most of the biggest oil companies in Canada, including several that are operating in the Athabasca Oil Sands a few hundred miles north, where thousands of miles of boreal forests are being disemboweled to satisfy the world's voracious thirst for oil. The process of converting the bitumen — the heavy, viscous tarlike substance that is mixed with the sand and water in the region — into refined oil takes huge amounts of energy, nearly all of which comes from fossil fuels. From a climate perspective, the oil sands (also called the tar sands) are an unmitigated disaster, and one of the chief reasons why greenhouse gas emissions in Canada are growing even faster than in the United States. The oil sands are also one of the chief reasons why Calgary is the poster child for our devil's bargain with fossil fuels: yes, you can have cheap oil and boomtown riches, but the price is environmental destruction and a superheated climate.

I rented a car (naturally) and drove out to the university on the northern edge of the city. In the distance, the Canadian Rockies cut a sawtooth horizon. The campus is uninspiring, with boxy Soviet-style architecture and acres of parking lots. Keith and I had arranged to meet near the entrance to one of the engineering buildings. He was a few minutes late, and while I waited, I couldn't help thinking about the contrast between this industrial-feeling campus and the sublimely beautiful Rocky Mountains in the distance. I wondered, If we can't even build a decent university campus, what are the chances we can successfully engineer the earth's climate? Maybe if I had been meeting Keith at, say, Stanford, where the buildings have some grace and beauty, I might have felt differently. But here I felt only dread. The fear, I realized, was not that we were going to geoengineer the planet. The fear was that we were going to do it badly.

Finally, Keith appeared, apologetic for running late. He was in his early forties, a tall, thin man with glasses and a strikingly slender face. He had a mountaineering backpack slung over his shoulder and was wearing brown corduroy pants and a white dress shirt.

He moved fast, talked fast, and had a quick and friendly smile. I immediately liked him.

We walked down a long, cold, empty corridor, chatting pleasantly about my trip. Then we came to a steel door, which he pulled open to reveal a room the size of a small warehouse. There were lifts and scaffolding everywhere, and a metal staircase leading down to the sunken floor of the central bay. There an enormous tube, about three feet wide and thirty feet tall—it looked like the stack of a small power plant—was hooked up to a tangle of wires, fans, pumps, and computers. Several students, dressed in blue coveralls and hardhats, crawled around the contraption, messing with wiring and keyboards.

Keith put on a hardhat and handed one to me. He pointed to the tower. "There's the beast," he said proudly. "It's not running right now—we're still doing some last-minute fiddling to get it working correctly."

We walked around to the back of the room, where the upper floor of the engineering bay allowed us to stand close to the top of the machine. He explained how the device works. "The process is very simple," he said. "The basic chemistry is stuff any high school student could understand." As Keith explained, lye (sodium hydroxide) is a strong base. When it is emitted from the machine and comes into contact with the air, it binds with the carbon dioxide, which is acidic, removing it from the atmosphere. The lye is transformed into sodium carbonate, a common compound that is also known as washing soda or soda ash, which is used in everything from laundry detergent to toothpaste.

Once the carbon is bound up in a solution of sodium carbonate, you need to figure out how to extract it, since sucking vast amounts of CO_2 out of the atmosphere to create mountains of toothpaste additive isn't very helpful (not to mention the vast quantities of lye required to keep this process going). So Keith's scrubber includes a second chemical reaction: mixing the sodium carbonate with lime causes the CO_2 to settle out in particles of calcium carbonate. The calcium carbonate is then thrown into a kiln and heated up, which causes it to separate into pure streams of CO_2 and quicklime (basically, just cooked lime). The CO_2 can be captured and sequestered underground, and the quicklime can be regenerated into lye and reused, creating an endless chemical loop that would, slowly but surely, pull CO_2 out of the atmosphere and cool the planet.

Compared to many chemical processes used by industry, this is simple stuff. The problem is, like everything else, it requires energy. And because of the low concentration of CO_2 in the air (only about 0.04 percent), building a machine to do it on a massive scale requires a lot of energy. But is there a way to optimize the process so that it becomes so cheap and efficient that you could afford to build enough towers—you'd need tens of thousands of them—to have a meaningful impact on the CO_2 content of the earth's atmosphere?

That, in essence, was the question Keith had set up this experiment to answer. He pointed to the top of the tower. "Inside, it's basically like a shower. The sodium hydroxide is sprayed out of a bunch of tiny nozzles, which rains down inside the tube. The trick is to maximize each droplet's contact with air, so that it absorbs as much CO_2 as possible." Then he pointed to what amounted to a large catch basin at the bottom of the tube. "We collect the sodium carbonate at the bottom. If this were the real world, we would heat the sodium carbonate up and separate out the CO_2. But since this is only a lab experiment, we're not doing that now. The main goal here is just to see what kind of efficiencies we can get out of the sprayer and get a better sense of what the economics of all this might be."

I marveled at the relative simplicity of this idea, and the high-powered imagination it took to build a contraption like this. Of course, by the time I arrived in Calgary, I had done some homework on Keith's project. I learned that the idea of scrubbing CO_2 out of the air is not revolutionary in itself; it's done in a variety of industrial settings, as well as in submarines to keep crews from asphyxiating on their own exhalations. I also knew that Keith was not the only scientist in North America working on a machine to pull CO_2 out of the atmosphere as a way to fight global warming. But his device was the only one, as far as I knew, that was actually up and running. And it had a charmingly homespun feel to it. The whole thing was built out of materials that looked as if they came from Home Depot—including the duct tape that held many of the parts together and the cheap, twisted wires that connected the computers to the tower. It reminded me of Leonardo da Vinci's famous drawings of men with mechanical wings and how the urge to fly was expressed through the crudeness of his imagination. The same was true here. What was remarkable was not the sophistication of the design but the fact that it existed at all.

Still, Keith's CO_2 scrubber felt an awful lot like a perpetual motion machine. Think about how our industrial infrastructure works: you have the largely human-built machines—from steam engines to Hummers—that have burned coal and oil and gas and filled the atmosphere with CO_2. Now, instead of shutting those machines down, you solve the problem by building another generation of machines to undo the ill effects of the first ones. And this new generation of machines, these carbon scrubbers, will require still more machines to power them, whether they be natural gas power plants or large-scale solar or wind power installations. As the British Royal Society noted in its 2009 report, capturing a significant amount of CO_2 directly out of the air could "require the creation of an industry that moves material on a scale as large as (if not larger than) that of current fossil fuel extraction, with the risk of substantial local environmental degradation and significant energy requirements."

The whole thing struck me as inspired but wildly impractical. Keith estimated that with a system like this, it would cost upwards of $500 for each ton of CO_2 the machine pulled out of the atmosphere. He was hopeful that, with some optimizing of engineering, that price might reasonably fall to $150 a ton or even lower. But that was still hugely expensive.

Keith was under no illusions about any of this. "It's a Russian tractor," he said. "There is nothing high-tech about this. But it works. It does what it's supposed to do—which is remove CO_2 from the air."

I asked him when he was going to turn it on. "Tomorrow, hopefully," he said. Then he went off to talk to his grad students. I wandered around for a few minutes, then noticed a small block of wood near the top of the tower. On it someone had scrawled in block letters PLANETARY COOLER.

Keith was born in Madison, Wisconsin, then moved to Ottawa when he was two. He was an only child and slow to learn to read. "I was worried about him," his mother, Deborah Gorham, told me. "Then we discovered that he was mildly dyslexic." When he was seven, his parents divorced. They both remained in Ottawa, and Keith spent his childhood bouncing between the two. His father, Anthony, was something of an environmental pioneer. He did groundbreaking work on the impacts of the pesticide DDT while he was a graduate student at the University of Wisconsin in the 1960s, then continued to explore the insidious effects of pesticides for the next thirty years as a researcher with the Canadian Wildlife Service. His mother stayed home with him when he was young, then went back to school to get a Ph.D. and spent the next several decades as a professor at Carleton University, where she started the women's history program.

Keith was a good student in high school but not extraordinary. He gravitated toward science in part because math was easier for him than reading. He spent a lot of time outdoors, exploring and camping in the woods near a small cabin his father owned outside Ottawa. He graduated from the University of Toronto with a degree in physics and, with his stepmother's help, got a summer job in the lab of Paul Corkum, a highly regarded laser physicist. Instead of going straight to graduate school, however, Keith took a year off, in part because he wanted to explore a place that had fascinated him since he was a kid: the Arctic.

In Ottawa, Keith lived down the street from Graham Rowley, one of the last of the great Arctic explorers and the first to chart Baffin Island, a remote island below Greenland, separated from Quebec by the Hudson Strait. Rowley's big Victorian house was stuffed with souvenirs from his trips to the far north—narwhal tusks, ivory carvings, and the like—with the occasional Inuk dropping by. Keith's father was also friends with Chuck Jonkel, who was leading polar bear studies in the high Arctic at the time. As Keith told me in a later email, "He would return with arctic char (a kind of super salmon) and marvelous stories of punching a polar bear in the nose that stuck his head through the window over the bed he was sleeping in." To Keith, the high Arctic was a place of romance and adventure, the last undomesticated corner of the planet, full of wild people and wild animals.

During his year off, between weekend rock climbing trips and a visit to Joshua Tree National Park in California, Keith took two journeys to the cold north. First, he camped alone in a remote region of Labrador for three weeks, near where Rowley had once camped. Then he spent four months living in a plywood shack on Dundas Island in the middle of the Arctic Archipelago, where he tracked walruses with renowned polar bear biologist Ian Stirling. The experience of living on Dundas and observing walruses had a profound effect on him, and he frequently cites it as one of the happiest times of his life. It was while he was on Dundas, in fact, that he learned (via shortwave radio) that he had been accepted to graduate school in physics at the Massachusetts Institute of Technology.

At MIT, Keith continued pursuing quantum physics—his thesis, which he completed in 1991, was on atom optics, a hot field at the time. (I downloaded a copy of his thesis before I left for Calgary, thinking I might read it on the plane. I finished exactly two paragraphs before becoming hopelessly lost in numbers and equations.) His adviser, David Pritchard, was a brilliant physicist known for grooming Nobel Prize winners. Keith himself seemed to be on that track. While at MIT, he built the first interferometer for atoms, which gained him instant visibility in the physics world. On the strength of this work, he was offered a cushy position at CalTech. He seemed destined to end up as another smart guy doing important work that six people in the world understood.

But while he was at MIT, Keith grew bored with physics. "For two hundred years, physics had been inward bound," he explained. "I doubt there are any more truly interesting discoveries to be found in that direction. Yeah, there are still unanswered questions about the exact interior constituents of nuclei—like whether we find a Higgs boson or not. But let's say we find a Higgs boson—does it give us any useful information about how the world works? At this point, I don't think there is anything you can find out in high-energy physics that is likely to have any big impact on anything we care about."

Keith found his attention drawn to more intractable problems out in the real world—especially global warming. In the late 1980s, climate change was just beginning to be discussed seriously at places like MIT. It was not just that the science of global warming was getting better, giving people a clearer sense of the dangers of unchecked greenhouse gas emissions. It was also part of the evolution of environmental consciousness beyond more visible issues such as air and water pollution. Keith's father's work on DDT, after all, had demonstrated that humans could have an impact on the planet that was far more damaging than dead fish in dirty rivers. This view was underscored in the mid-1980s when scientists found a hole in the ozone layer in the earth's atmosphere. Chlorofluorocarbons (CFCs), compounds used in aerosol spray cans and refrigerants, were thought to be relatively harmless substances. In fact, they were destroying the layer of the atmosphere that filters out the sun's ultraviolet rays, potentially putting millions of people at risk for skin cancer and related diseases. Had the ozone hole grown undetected for a few more decades, the damage to the atmosphere could have been catastrophic. But for Keith, the important point was that it was scientists—including Paul Crutzen, the atmospheric chemist who later piqued my interest in geoengineering—who had detected and diagnosed the ozone hole. They weren't messing around with lasers in labs. They were saving the world.

From Keith's perspective, the ozone hole and global warming were very different phenomena, but there were obvious parallels, too: both involved an invisible compound that human beings were emitting into the atmosphere, both had profound consequences for the human race, and both—presumably—were problems that could be fixed. For ozone depletion, the solution was fairly simple: ban CFCs and figure out another way to propel aerosols out of cans and keep refrigerators cold. But for global warming, the fix was not so obvious. You couldn't ban greenhouse gases without shutting down the global economy. Indeed, given the close correlation between CO_2 emissions and economic power, it was clear that even getting nations to agree to limit pollution would be a tough task.

Keith was fascinated, drawn not only to the complexity of the problem but also to the moral urgency of solving it. While he was working on his thesis, he began meeting informally with other students interested in global warming, including Ted Parson, now a law professor at the University of Michigan, and David Victor, now the director of the Laboratory on International Law and Regulation at the University of California, San Diego. They talked about the problem from every angle, from the promise of renewable energy to the difficulty of enforcing an international agreement to cut emissions. (This was five years or so before the 1992 Earth Summit in Rio de Janeiro, which marked the beginning of the United Nations process that eventually led to the Kyoto Protocol, the first international agreement to limit greenhouse gas pollution.) Among other things, they wondered, was it possible to design an energy system that didn't emit greenhouse gases? Or was there a way to deal with warming that didn't require rebuilding the energy infrastructure of the civilized world?

When Keith read the scientific literature on global warming, he discovered that the very first presidential report on the problem, which landed on the desk of President Lyndon Johnson in 1965, didn't talk about cutting greenhouse gas emissions at all. In fact, the only solutions the report proposed were what we'd now call geoengineering fixes, such as putting floating particles in the ocean to reflect sunlight or dumping iron into the ocean to stimulate plankton blooms and suck up CO_2. This surprised Keith, in part because by the late 1980s, these solutions had vanished from the main political debate about global warming, which was already focused on pushing forward an international agreement to limit greenhouse gas pollution. Why was no one talking about geoengineering anymore? "I figured it was like sex," Keith recalled. "If nobody was discussing it openly, there had to be something interesting there." Keith began batting around geoengineering ideas with his friends at MIT. Nobody thought he was crazy. It was an engineering school, after all. The only criteria were, was it economically feasible, and if so, would it work?

When it finally came time to make a decision about his future, Keith spent a lot of time talking it over with his father. "David was still thinking about going out to CalTech to pursue a career in laser physics," Anthony Keith recalled. "I said, 'Well, what are the social consequences of this information? Where do we go with it? What do we do with it? Are you going to follow an academic path, or do something about the world?'"

Not surprisingly, Keith decided to turn down the posh job at CalTech and instead took a position at Carnegie Mellon University in Pittsburgh. In effect, it was a decision to give up mainstream physics and devote his career to energy and global warming. CMU's engineering school is noted for its philosophy of applying engineering know-how to complex policy questions. Not long after he arrived, Keith began working on systems to capture CO_2 from coal plants and sequester it underground—an important technology that, in theory, would allow us to burn coal without trashing the atmosphere. But Keith's fascination with geoengineering remained. Shortly after he arrived at CMU, he coauthored his first paper on the subject—titled, appropriately enough, "A Serious Look at Geoengineering."

The article was published in *Eos,* the journal of the American Geophysical Union, a fairly prestigious scientific publication. But the paper didn't exactly spark a revolution, in part because Keith was a young scientist with a slim publication history, and in part because—unbeknownst to Keith—the prestigious National Academy of Sciences had also been working on a geoengineering study, which was released about the same time as Keith's paper. The NAS report was the product of a number of top-level scientists working together to come up with a definitive statement about the strengths and weaknesses of various geoengineering strategies. Keith's paper was more or less the work of one man's curiosity and passion. (Keith's coauthor, Hadi Dowlatabadi, a more senior professor at CMU, has described his own contribution as "modest.")

In just a few pages, Keith's paper laid out the logic for taking geoengineering seriously, which was (a) that climate change might happen faster than we expect and (b) that the world was unlikely to cut emissions fast enough to stave off abrupt climate change. The paper offered a rundown of various ways we might remove CO_2 from the atmosphere, from planting trees to capturing the gas from smokestacks and burying it in the deep ocean (curiously, air capture was not mentioned), as well as albedo engineering schemes, such as injecting particles into the stratosphere and launching mirrors into space. It briefly covered the engineering challenges of each one, offering a short analysis of potential side effects.

For me, the key point in Keith's paper was how inexpensive some of these geoengineering strategies could be (the NAS study had come to a similar conclusion). He calculated, for example, that enough sulfur dioxide particles could be injected into the atmosphere to counter the impact of a doubling of CO_2 emissions from preindustrial levels for about $30 billion a year, or the equivalent of 0.7 cents per ton of CO_2. This is, by any standard, peanuts. (In the United States, policymakers assume that a carbon tax or trading scheme—if and when it is finally implemented—will initially levy a price on CO_2 emissions of $10 to $30 per ton.) Others have since revised the estimated cost of spraying particles into the atmosphere to be three times higher than Keith's 1992 estimate, but in the grand scheme of things, that's still a bargain. Keith's paper confirmed that whatever the limiting factors of geoengineering might be—ethical concerns, fear of environmental side effects—cost was not one of them.

This implied that unlike, say, building a nuclear bomb, which is complex and expensive, shooting particles into the atmosphere would be so easy and inexpensive that it could be done by virtually any half-prosperous nation, including half-prosperous nations with evil intents. But it might also be done by a well-intentioned billionaire with a fleet of high-altitude airplanes. (Richard Branson's Virgin Earth Challenge!)

Of course, anyone who has been engaged in a thoughtful discussion about how to confront global warming should be immediately suspicious of any talk of "cheap" solutions. It reminds me, in fact, of coal industry executives who are always touting the low cost of coal-fired electricity. If you measure the cost of power only on a per-kilowatt basis, dirty coal plants do generate cheap power. But if you include the real costs of burning coal—the blasted mountains where the coal is mined, the hospital visits of children with asthma, the rapidly warming planet—coal is not cheap at all. In fact, you could just as easily argue that it is the most expensive form of energy we have.

I quickly understood that it's the very cheapness—or, rather, the *perceived* cheapness—of some kinds of geoengineering that makes them so dangerous. Who can resist a cheap fix? Certainly not fearful politicians, governing in unstable times, with troubled economies and social turmoil.

Keith recognized these risks, too. As he put it succinctly in his paper:

> Important uncertainties in the political implications of geoengineering include *Sovereignty:* Who has the ability to deploy such a scheme? *Equity:* How are costs (direct and indirect) distributed relative to benefits? *Liability:* Creators of a geoengineering system will be blamed for an obvious failure. Would they be de facto liable for natural climate fluctuations? *Security:* Might geoengineer-

ing security systems (for example, solar shields) be construed as offensive weapons?

Reading this, I found it hard not to draw loose parallels with the rise of nuclear weapons and the ways in which a new technology fundamentally changes how we think about the world and our relationship with both technology and nature. It occurred to me that Keith was precisely the kind of elite scientist who would have been toiling away at Los Alamos—a bright young physicist drawn to real-world problems, a man who believed in progress and the ability of humankind to manage complex tools. At least that was how it seemed to me at the time. But what interested me most was not the technological questions about geoengineering, but the human ones. What kind of person dreams about engineering the entire planet? And, given how much is at stake, can we trust him?

After about an hour of messing around with the Planetary Cooler, we headed to a nearby student café, where we grabbed some coffee and had our first extended conversation about God, global warming, and geoengineering. Actually, I'm not sure God was part of the discussion, but given the scale of the subjects we addressed, he may as well have been.

The first thing I learned about Keith is that, like any rational person, he believes strongly that the best way to tackle global warming is to cut greenhouse gas pollution. And it wouldn't even be all that hard or that expensive to do so if we really set our minds to it. "The cost of reinventing our energy infrastructure would be, in any meaningful sense, trivial," he said. He estimated that for about 2 percent of gross domestic product, which is about what America spends on the military and far less than what it spends on health care, we could transition fairly quickly to a zero-carbon economy. So why don't we just do it? He smiled knowingly. "That is the question, isn't it?"

The second thing I learned is that Keith loves hardware, big and small. He debates the merits of Macs versus PCs and nukes versus coal with equal aplomb. There is something of the boy with the Erector set in how he thinks about energy, looking at it from a hyperrational and systematic approach, as well the more exuberant "Look what I can build!" Perhaps it goes without saying that a person who is interested in geoengineering the planet would be interested in cool hardware, but the connection had not occurred to me before I met Keith. He underscored this initial impression for me a year or so later, when I met him after a conference at Harvard University. He and I had been talking a lot about how particles might be engineered to better reflect sunlight in the stratosphere. On this particular evening, he sat down beside me and joked, "I think I've figured out how to build tiny flying saucers." By that he meant tiny particles that would self-levitate into the stratosphere. (It didn't help that when he mentioned this idea, he was dressed entirely in black and looked like someone who had wandered off the set of *Austin Powers: International Man of Mystery.*)

The third thing I learned is that Keith has a broad contrarian streak in him. He likes to challenge conventional wisdom, such as my suggestion that solar is the ultimate answer to the world's energy problems. "Have you done the numbers on solar scale-up?" he asked pointedly, implying that solar power is too costly to solve our energy problems anytime soon. He has other un–politically correct views, too. He coauthored a study suggesting that big wind farms can change evaporation rates in the soil, altering weather patterns in the region (a view that didn't exactly endear him to wind power advocates). He thinks environmental opposition to nuclear power is overblown. And you won't hear him touting the virtues of energy efficiency too often: "You buy a car that gets better gas mileage, and what do you tend to do? Drive it more."

But the most interesting thing he said during that first meeting was this: "I'm not sure that global warming is such a threat to human civilization. I think we have to be honest with ourselves—there will be winners and losers. Some places will experience more productivity. Some people will enjoy warmer weather. This is not to deny or minimize the suffering and hardship that others will experience, especially in poor countries. But the fact is, human beings are a remarkably adaptable species. And I believe that, by and large, people will adapt to the changing climate. If it is just the human race you're worried about, I'm not sure global warming is such a big problem."

"So then why bother with geoengineering?" I asked.

"Two reasons," he said bluntly. "First, I might be wrong. We don't know exactly how big the changes will be or how fast they will come. And we don't know how people will react."

He paused, letting that sink in.

"Second, geoengineering may be the only tool we have to save certain ecosystems, like the Arctic."

"So geoengineering is a conservation tool?"

"In a sense, yes. If you value a place like the Arctic, if you decide that it has some meaning, some connection to our common human heritage, and you want to preserve it, it's not exactly clear how you would do that other than geoengineering. In theory, you could inject particles into the stratosphere above the Arctic and stabilize the ice in the region—or if you wanted to, even grow some of the ice back. There would be consequences and side effects. I'm not saying this is a perfect solution. But as far as I can see, it's the only tool we have."

Later that evening, we checked back in with the students working on the Planetary Cooler, then walked out into the parking lot. Although I generally resist making judgments about people based on the kind of car they drive, it was satisfying to see that Keith owned a Toyota Prius. And not just any Prius—his was filthy, beat-up, driven hard.

I followed him across the city in my rental car. My initial impressions of Calgary were confirmed—the commercial sections were one big strip mall. But then we twisted and turned our way through some residential neighborhoods, winding up in a section of the city called Elbow Park. It was full of modest old homes and big trees. Many of those homes were becoming immodest new homes—teardown mania was rampaging through the neighborhood. Keith pulled up at the curb of a cool, 1940s-style house near the end of the block. Although we were in the heart of the city, the setting felt wild. The Elbow River runs through Keith's backyard, with a high hill rising up behind it, blocking the city lights.

We stepped inside Keith's house, taking off our shoes and leaving them near the door. He dropped his backpack and yelled out to his kids—his son, Alex, who was nine, shouted down from upstairs. His daughter, Sarah, age seven, looked up from a book she was reading in the kitchen and mumbled hello. The kitchen was spacious and bright, but the general feeling of the place was post–graduate school family chaos, with lots of old books and well-worn furniture. He kissed his wife, Susan, an athletic-looking woman with blond hair, then they spent a few minutes going over the news of the day. All in all, it was a scene right out of *Leave It to Beaver,* except that instead of working in an office, Dad was engineering the planet.

As we ate dinner, I couldn't help but think that scientists like Keith are likely to be the superheroes of the geoengineering era, building the tools and technologies that will shape the world we live in. But it also struck me that he is a man with a family, a house, a car, a normal life cluttered with bills and trivial errands, and, most of all, a character that is shaped by personal history and random genetics (subject to hubris as well as humility) and a mind that is brilliantly analytical as well as capable of silly mistakes. In other words, he is a human being. Are we—and by this I mean not just Keith but all of us in the PlayStation aristocracy, the Ph.D.'s of progress—ready to exert our influence over the planet in this way? At dinner that night, I thought of all the troubles the best and the brightest have gotten us into in the past century. It occurred to me that, again like nuclear weapons, one of the distinctive qualities of geoengineering is that it gives us the chance to screw things up on a truly grand scale.

After dinner, Keith and I sat on the couch in the living room and talked some more. I asked him about one of the most rational and straightforward objections to geoengineering, which is that our climate is a highly complex system—if you start pushing it around, you never know what will happen. This objection is most often raised against albedo engineering schemes, such as throwing particles into the stratosphere or brightening clouds, which could quickly and dramatically change heat distribution on the planet. People often use the familiar analogy of a butterfly flapping its wings in China and causing a hurricane in the Gulf of Mexico to describe this phenomenon. And certainly, this idea makes intuitive sense. The fact that even the best models can't accurately predict next week's weather is evidence of just how dynamic the system is. So given these uncertainties, it seems downright foolish to think that we could ever understand the climate system well enough to control it. Or so I suggested to Keith that night after dinner.

"Not necessarily!" he said, with what I now recognized as contrarian gusto. "Predicting weather and predicting climate are two different things. A system can be quite chaotic on a local or regional level, while much steadier on a global scale." Keith explained that weather is what scientists call an initial value problem, while climate is a boundary value problem. He used the analogy of an airplane wing to explain the difference. On the one hand, Keith said, it's virtually impossible for engineers to predict the exact swirls of turbulence created by a wing as it moves through the air. Any error in initial calculations of the swirling air will quickly lead to radically different end points. (In a similar way, if you drop two leaves into a rapidly flowing stream at the same point, they will get caught up in different currents and eddies and, after a short distance, end up in quite different places in the river.) Weather is like that—an initial value problem. On the other hand, climate is a boundary value problem, like knowing at what speed and angle the plane will stall. "Engineers can calculate that very precisely," Keith said, because it is dependent not on tiny swirls of air, but on basic calculations of air pressure and lift. "Predicting the climate is more like that—you don't have to know the exact swirls of atmospheric flow a century in advance to know that increasing solar absorption makes the planet warmer."

"It's crucial to communicate this distinction," Keith emphasized. "Otherwise, whenever you talk to the general public about global warming, the natural reaction is to say, *How the hell can you jokers claim to predict the climate a century in the future when you cannot predict the weather two weeks out?*"

But in Keith's view, even chaotic systems like weather are not necessarily impossible to control. "Let me show you something," he said, as he jumped up to get his laptop—a Lenovo ThinkPad. He pulled up a picture of a very odd-looking military jet. It looked like the wings were on backward—instead of swooping toward the rear of the aircraft, like the wings of every plane I've ever seen, they were bent forward. "This is an X-29," Keith told me, "an experimental plane the military built not so long ago. You look at it, and you think it's an unstable design—a classic example of an unstable system. And if you try to fly it the same way you fly a regular airplane—it *is* unstable. You will crash. But if you understand the rules of this new system and fly the plane in a different way, it turns out that, at certain speeds, it is actually *more* maneuverable than a traditional aircraft."

"Are you saying the weather is like an X-29?"

"No, I'm not. I'm just saying that even highly complex systems are not *necessarily* beyond our control."

I've proposed this idea to a number of scientists—some agree, some don't. Those who disagree point out that "control" of a system like the earth's climate is complicated in itself and that how you think about it really depends on what your goals are. If your goal is to maintain a certain global mean temperature, that's a relatively straightforward engineering problem. But if your goal is to maintain a steady level of annual rainfall over, say, Phoenix, then that is another thing entirely. After all, temperature and precipitation are two very different things, driven by different (but not unrelated) forces. And just because you can cool the overall temperature of the earth, it doesn't mean the impact will be uniform over all the earth. As Keith later explained to me, "It's absolutely accurate to think of geoengineering as building a temperature control knob for the planet. But in that metaphor, the planet is really more like a big old house. When you set the thermostat, it doesn't heat all the rooms equally. Some rooms will be cold and drafty, some too warm. You'll discover new leaks and drafts."

Other geoengineers use different analogies when they talk about controlling the earth's climate system. Stephen Salter, a Scottish inventor who has designed cloud-making machines, likes to compare learning to manipulate the climate to learning to ride a bicycle. "Imagine you had never seen a bicycle before, and someone handed you one and asked you to ride it," he told me. "You would have no idea what to do. If you tried to sit on it, you would fall over. But after you played around with it for a while, you would figure out what the handlebars are for, what the pedals are for, how the brakes work. And you would figure out that once you got moving, the bike was pretty stable. After a while, you would get so good on the bike that you might think it was a pretty elegant machine. Well, maybe the climate system is like that. Maybe we just haven't found the handlebars yet."

The longer Keith and I talked, the more our conversation kept circling back to the Arctic. From a scientific point of view, it's no surprise. In the Arctic, the situation is dire. Even in a modest warming scenario, it is a goner. The only question is how long it will take to disappear. I've met plenty of environmentalists who bemoan the fate of the polar bears, and I've met climate researchers who make a strong argument about the importance of the Arctic in the earth's overall climate system. But until I met Keith, I'd never met a scientist who had emotional feelings for the place, who regarded it with something like, well, *love*.

But clearly Keith did. That night, he mentioned Graham Rowley, the great Arctic explorer who lived in his neighborhood in Ottawa. When Keith was in his thirties, he and a friend cross-country skied the entire length of Baffin Island—some three hundred miles—essentially following in the footsteps of Rowley. He showed me a pair of small polar bears carved out of walrus tusks that had been given to him by an Inuk named Aipilik, who had been a friend of Rowley's. He talked about the four months he had spent tracking walruses on the Arctic Archipelago.

In fact, it was on this trip that Keith learned something about how deeply human beings are connected with nature—a lesson, he believes, with direct relevance to the work he is doing now. "We were listening to walruses underwater with hydrophones, which were programmed to pick up the sounds of walruses," Keith told me. "But what amazed me was that one of the techs could pick up the walrus sounds in the air around camp, too. Not because his hearing was better—once he pointed out the sounds to me, I could hear them, too—but because he had spent so much time around walruses that he was acutely attuned to them. To me, it really spoke to the deep kinship we have even with walruses. We humans can talk, but there is a hell of a lot we have in common with large mammals."

He explained: "What I mean is, if the Zorthians from Planet Zorthia came to earth, they could do quantum physics, but they couldn't understand the walrus as we do. Our feeling for nature is based on our genetic kinship—what the biologist Edward O. Wilson calls "biophilia." We feel good when we sit by a campfire, run through the woods, or interact with a big mammal because it touches something deep within us. Maybe we can figure out a way to engineer our food supplies, our drinking water supplies, even our climate. But we still evolved—our brains evolved—in the natural world. And preserving a link to that world is important—not just for me personally, but also as a larger anchor for our society, even as we gradually engineer a social order that largely is cut off from nature."

As I learned during subsequent conversations, Keith's biggest fear about geoengineering has little to do with control systems or changing precipitation patterns. It has to do with what happens when we sever that primal link with nature that has shaped human evolution. "This is why geoengineering is so dangerous, and why we need to be careful about how we pursue this," Keith told me over dinner one night in Washington, D.C. "It's not the end of nature — but it is the end of wildness — or at least our idea of wildness. It means consciously admitting we're living on a managed planet. It may be that geoengineering can help save the Arctic. But it won't be the same Arctic we have today. It will be a museum piece, a place for the elites to go someday and remember what the real Arctic used to be like. The fact is, whether we want to admit it or not, we're living in a zoo. And we're both the animals and the zookeepers now."

Back in Calgary, I met Keith on campus the next morning for the start-up of the Planetary Cooler. We donned our hardhats and descended to the floor of the engineering bay, where one of Keith's postdocs was working on a computer. This was not, I was disappointed to learn, the first time they had turned the Cooler on. In fact, it had been running the week before, but Keith had shut it down to make some adjustments in the sprayer and the circulating-fluid system. Still, there was some tension in the engineering bay that morning, some sense that they really wanted to get it right. "Next week," Keith said, "we take it apart."

"Are you going to donate it to the Smithsonian?" I asked, half-joking.

He smiled. "Maybe the next one we build."

As I waited for them to get the Cooler ready to go, I thought about a photo I had once seen of the first electronic transistor (or one of the first), which was built at Bell Labs in the late 1940s. It was a rough, kludged-together thing, with fat wires and a big wedge-shaped hunk of insulating material on a heavy metal base. But it worked, and, conceptually, it was a lot simpler and more reliable than the vacuum tubes it replaced. The story of the transistor has been told many times, but it is nevertheless mind-boggling to hold an iPhone in your hand today and think of how many technologies have evolved from that one awkward invention. Of course, no one could have foreseen such progress — just as no one could foresee what might come of that duct-taped machine Keith had thrown together in the engineering bay. Right now, capturing CO_2 directly from the air remains an expensive, impractical idea. But given our climate problems, and given how much CO_2 we have already pumped into the atmosphere, it is not unimaginable that we will someday figure out how to build what amounts to an iron lung for the planet.

Still, turning on the Cooler felt like a big moment in itself. It felt like a leap across a big divide in the development of the modern world—from making machines that dump CO_2 into the atmosphere to making machines that scrub CO_2 out of it. From making machines whose primary goal is to serve humanity to making machines whose primary goal is to repair and maintain the planet. Primitive as it was, the Cooler seemed like a small gesture of benevolence.

"You ready?" Keith asked one of his postdocs.

He nodded.

"Let's see what happens," Keith said.

There was hardly any sound, nothing dramatic. Just the quiet whir of fans pushing air through the Cooler; the gentle rain of sodium hydroxide falling through the tower; the hum of a pump.

"Is it working?" I asked.

Keith pointed to a graph on the computer—two wiggly lines, scrolling out slowly, one above the other. "The top line is the amount of CO_2 in the air coming in at the top of the tower," he said. "It's about 340 parts per million of CO_2." Then he pointed to the lower line: "And that's the amount of CO_2 in the air coming out the bottom. It's about 335 parts per million. So it's about five parts per million lower."

He shrugged modestly. "It's a start."

THREE

God's Machine

IT IS NOT ACCURATE to say that David Keith lives in the future, because he is as rooted in the here and now as any of us. But he does live in the world of possibility, of exploration, of imagining what could be. Geoengineering itself is a little bit like that, since it is still more of a technological fantasy than a practical reality. In certain ways, the whole idea could be said to be a spinoff from the invention of the computer chip, since it is only the rise of computers that has given us the ability to begin assembling a dynamic picture of the planet, exposing the levers and wheels that spin and clank and make this big rock we live on behave as it does.

When I returned from Calgary, however, I decided to put the future aside and spend a little time thinking about the past. What kind of world had this dream of manipulating the climate emerged from? Who were the pioneers? What might we learn from them?

The first thing I discovered was that although geoengineering may feel like a shiny new idea, it's not. In fact, it is deeply intertwined with the history of climate science itself. In the 1890s, Swedish physicist Svante Arrhenius was the first to understand the role that CO_2 plays in regulating the earth's temperature. After many months of tedious pen-on-paper calculations—a process that was so all-absorbing that his beautiful young wife abandoned him while he was in the midst of it—Arrhenius calculated that a doubling of CO_2 levels in the atmosphere would lead to a rise in temperature of about 7 to 9 degrees Fahrenheit. That's remarkably close to current estimates and underscores just how rock solid the fundamentals of global warming really are. Because Arrhenius's work is so basic to our current understanding of the greenhouse effect, he is rightly regarded as the godfather of modern climate science.

But unlike many modern climate scientists, Arrhenius was not worried that we would burn too much fossil fuel and cook the planet. In fact, Arrhenius, who often admitted that he was tired of the long, cold Swedish winters, saw coal—the dominant fossil fuel of his time—as a tool for manipulating the earth's climate. As he wrote in 1906, by burning coal we might create a planet with "more equitable and better climates, especially as regards the colder regions of the earth," and this warmer climate would improve food production "for the benefit of rapidly propagating mankind." Arrhenius's friend and fellow researcher Nils Eckholm agreed, but argued that the process of heating up the earth by burning coal for industrial uses was taking too long. He recommended just setting fire to shallow coal seams, which would release more CO_2 into the atmosphere and turn the planet into a palm tree paradise.

This dream goes back a lot farther than the early days of global warming science, of course. Humans have been interested in manipulating the climate ever since we climbed down out of the trees—although weather, not climate per se, was the main issue. Climate is usually defined as the long-term average of weather and is a relatively recent concept.

Until the Enlightenment, the heavens were considered the playground of the gods. Controlling the weather was something you achieved by coddling, flattering, or tricking the supreme beings to hold off on storms and droughts. In *The Odyssey*, Homer refers to Zeus as "the cloud gatherer," and the god's mood is the subject of much worrying and seduction, lest he get upset and throw thunder-bolts from the sky. In Norse mythology, the skies were dominated by Thor, whose hammer caused thunder. In China, dragons kicked up storms and controlled the waters. In many cultures, the sacrifice of animals, especially black ones, was a popular way to ensure a salubrious climate. The Hopis danced with rattlesnakes in their mouths, believing that the snakes could carry the dancers' prayers for rain to the other world. (To the Hopis, novelist D. H. Lawrence wrote, "everything is alive, though not personally so. Thunder is neither Thor nor Zeus. Thunder is the vast living thunder asserting itself like some incomprehensible monster, or some huge reptile-bird of the pristine cosmos.") Often these rituals had a sexual subtext. The Dieri people of central Australia believed that the foreskins taken from boys at circumcision had great rain-producing power. At the turn of the twentieth century, temperance movement activists tried to stop the Hopi snake dance because it was said to include "simulations of group fornication."

Often the gods used rain, wind, and storms to cleanse the earth of human corruption. Noah's flood is the most obvious example. In other cases, it's the foolishness of the gods themselves that causes problems. In Greek mythology, the scorching heat of Africa and the burnt skin of its inhabitants were attributed to Phaëthon, an offspring of the sun god Helios. Phaëthon won a wager with his father, which allowed him to drive the sun chariot across the sky. In what climate scientist Kerry Emanuel called a "primal environmental catastrophe," Phaëthon was struck down by Zeus, frying the earth and getting killed in the process.

In the West, faith that gods or angels were controlling the earth's climate ended about the time the apple fell on Isaac Newton's head. As Alexander Pope put it:

> Nature and Nature's laws lay hid in night:
> God said, "Let Newton be!" and all was light.

One of the primary impulses of the Enlightenment was to bring order to the chaos of nature, to break it down into mechanical parts.

"Science was the engine of the Enlightenment," wrote biologist Edward O. Wilson.

> The more scientifically disposed of the Enlightenment authors agreed that the cosmos is an orderly material existence governed by exact laws. It can be broken down into entities that can be measured and arranged in hierarchies, such as societies, which are made up of persons, whose brains consist of nerves, which in turn are composed of atoms. In principle at least, the atoms can be reassembled into nerves, the nerves into brains, the persons into societies, with the whole understood as a system of mechanisms and forces. If you still insist on a divine intervention, continued the Enlightenment philosophers, think of the world as God's machine.

It was a machine they struggled to make sense of. In 1735, botanist Carolus Linnaeus published *Systema Naturae,* the first attempt to classify plant and animal life into groups and subgroups. In 1751, the first edition of the thirty-five-volume *Encyclopédie,* the explicit goal of which was to catalog all human knowledge, was published in France. After mathematician Johannes Kepler discovered the laws of planetary motion, seventeenth-century astronomers began mapping the Milky Way and charting asteroids. In 1803, Luke Howard, an amateur meteorologist in London, tried to do for the sky what Linnaeus had done for plants and animals, publishing his "Essay on the Modifications of Clouds," which broke clouds down into four categories: cumulus, Latin for "heap"; stratus, Latin for "layer"; cirrus, Latin for "curl of hair"; and nimbus, Latin for "rainstorm."

Early Americans were fairly obsessed with deconstructing the earth's climate system. Ben Franklin was among the first to figure out that storms have structure and behave in predictable ways. In 1724, on his first crossing of the Atlantic, Franklin and his companion regularly checked and recorded sea temperature along the way, startled to find that the water temperature was 20 degrees Fahrenheit warmer in the Gulf Stream. With this discovery, Franklin was one of the first to intuit a link between weather and ocean temperature changes. (The map he drew of the Gulf Stream is still considered accurate.) For more than fifty years, Thomas Jefferson recorded climate and weather information in notebooks, hoping to build what he called "the indexes of climate"—temperature, prevailing winds, precipitation, and related biological events such as the flowering of plants and the migration of birds—which was intended to form the foundation of a reliable theory of weather and climate.

A key figure in developing these early theories was James Pollard Espy. Espy, who was born in 1785, spent his youth on the western frontier, teaching school and practicing law in Ohio. In 1817, he moved to Philadelphia, where he devoted himself to meteorological research, eventually becoming the first meteorologist employed by the U.S. government. Espy helped equip weather observers in each of Pennsylvania's counties with barometers, thermometers, and other standard instruments to provide a more coherent picture of the weather, especially the development of storms.

This was one of the big questions of the nineteenth century: if storms didn't come from Zeus's thunderbolt or Thor's hammer, where exactly did they come from? Espy, drawing an analogy with the dominant mechanical device of his time, the steam engine, viewed the atmosphere as a giant heat engine. "According to [Espy's] thermal theory of storms, all atmospheric disturbances, including thunderstorms, hurricanes, and winter storms, are driven by 'steam power,'" wrote one science historian. "Heated by the sun, a column of air rises, allowing the surrounding air to rush in. As the heated air ascends, it cools and its moisture condenses, releasing its latent heat (this is the 'steam') and producing rain, hail, or snow." Espy's theory was a fundamental breakthrough, the first accurate sketch of the thermal dynamics of weather.

But Espy took it further, believing that this same insight could be used to modify weather. Espy noted how rainfall was higher over volcanoes, fuel-burning cities, and accidental fires, and he wondered whether it might be possible to create rain by artificially raising a large, well-directed column of warm air. Espy called on the U.S. government to build a network of vast, forty-acre bonfires, stretching some six hundred to seven hundred miles north to south—perhaps the first proposal to alter the weather on anything like a planetary (or at least a continental) scale. The pyre complex would be built somewhere in the Midwest, where it would burn throughout the summer. The result, he predicted, would be "a rain of great length, north and south," eliminating all droughts, increasing "the proceeds of agriculture," and enhancing "the health and happiness of the citizens." Despite Espy's reputation, there wasn't much interest in his climate manipulation scheme. One contemporary assessment of it: "Magnificent Humbug."

Humbug or not, the leap from describing a mechanical system to manipulating it was inevitable. If storms (and, by extension, the climate) were created by physical forces, not gods or beasts, then it followed that storms (and the climate) might someday be influenced by people.

French science fiction writer Jules Verne was among the first to understand the enormous consequences of this idea. In his 1889 novel *The Purchase of the North Pole* (a sequel to *From the Earth to the Moon*), Verne tells the tale of greedy business people who want to melt the North Pole so that they can mine the huge coal deposits that are supposedly beneath the ice. Their method: fire a giant cannon, the recoil of which will shift the earth on its axis, tilting the North Pole closer to the sun. In the logic of the novel, this would change the angle at which the sunlight hit the earth, melting the ice at the pole and allowing the business people access to the coal deposits underneath. Of course, bringing more sunlight to the pole would mean less sunlight in other places—but as long as they get their coal, so what if the Amazon rain forest freezes? As a parable of humankind's growing power over nature and our ruthless exploitation of the planet, it works pretty well. But what's really interesting is that the key figure in this dastardly plot is a mathematician who holes up in a room for weeks at a time, working out the precise mathematical formula for the placement and angle of the cannon. The fate of the world—and the future climate—depends on his calculations. Forget the angry gods, Verne suggests. Geeks rule.

Verne's mathematician bears a striking resemblance to Svante Arrhenius, struggling through pages and pages of calculations on the heat-trapping qualities of CO_2 in the atmosphere. But at about the same time as these early climate nerds were crunching numbers, a generation of pseudoscientific weather hackers emerged out of the American Midwest, eager to exploit this newfound faith in a mechanical universe. And they weren't waiting for precise calculations.

Rainmakers were the mutant offspring of P. T. Barnum, Harry Houdini, and Thomas Edison. They rode into desperate, dusty towns, promising to conjure water from the sky—for a price. (To give themselves an air of scientific legitimacy, some referred to themselves as "pluviculturists.") Most were con men out to make a quick buck off other people's suffering. For example, the grandly named Dr. George Ambrosius Immanuel Morrison Sykes, in addition to making rain, announced himself as a "minister of Zoroastrianism" and vowed to wage war on prohibition, vivisection, free love, and internationalism, among other seemingly unsavory endeavors. To make rain, he used a Rube Goldberg machine that included, according to one newspaper account, "radio apparatus, antenna, lightning coordinate grounds, and cloud attractors" to create "thermo-magnetic field impulses."

"I merely assemble rain-bearing clouds over a specified area," Sykes boasted, "and then I turn my engines on them. The earth's magnetic field fairly eats out of my hand."

Some rainmakers believed they were explorers on the cutting edge of science, unraveling the secrets of the skies. One common theory involved electricity. Rainmakers known as "putter-inners" would charge the atmosphere electrically to bring forth moisture. Those known as "taker-outers" would remove current from the air as a sure means of getting rain. "Boomers," or cloud bombers, hy-

pothesized that the smoke and vapor of burning gunpowder produced a chemical reaction favorable to the formation of raindrops
in the atmosphere. Cloud bombers often set up elaborate networks
of explosives, including gas-filled balloons and specially prepared
artillery shells, in the region where they hoped to produce rain.
One infamous cloud bomber explained to a reporter that his explosions—fueled by a special brew of oxygen, hydrogen, potassium
chlorate, and petroleum—produced "something in the nature of a
vortex or of a momentary cavern, into which the condensed moisture is drawn from afar or falls." The effect of this, he believed, "may
squeeze the water out of the air like a sponge."

In part, rainmakers were a manifestation of a hyper-confident
nation that felt, as New York senator Chauncey Depew famously
declared in 1900, "four hundred percent bigger" than the generation before. That hyper-confidence was boosted by a series of scientific and engineering breakthroughs that makes today's digital
revolution seem positively modest. Telephones, sound recordings,
dynamos, light bulbs, typewriters, chemical pulp, reinforced concrete—all these were invented between 1865 and 1880. The aston-
ishing 1880s—"the most inventive decade in history," according
to science historian Vaclav Smil—brought reliable and afford-
able electric lights, electric power plants, electric motors and trains,
transformers, steam turbines, the gramophone, photography, the
internal combustion engine, motorcycles, automobiles, and steel-
skeleton skyscrapers. Coming in the next few decades were x-rays,
movies, airplanes, tractors, radio broadcasts, vacuum tubes, neon
lights, stainless steel, and air conditioning.

If all that was possible, who was to say you couldn't make rain
fall from the sky? "After the telephone, after the automobile and
the aeroplane, came the harvesting of the desert," wrote Jonathan
Raban in *Bad Land*, an account of the settlement of the northern
plains. "They could watch the miracle happening as they walked
home from the schoolhouse: the ripening corn; the cows with their
calves; the electrically induced rain. It was like the wireless. It was
like five loaves and two fishes."

But the popularity of rainmakers was also a manifestation of the hard luck and desperation that followed the settlement of the West. The real engine of western settlement was not the locomotive — it was the idea, popularized by Charles Dana Wilbur, an advance man for the Burlington Northern Railroad, that "the rain follows the plow." So people moved into the arid West, pursuing free land and taking it on faith that, once they broke ground, the rain would come, and they could feed their families. As Wilbur put it, "[The plow] is the instrument which separates civilization from savagery; and converts a desert into a farm or garden." How exactly this process worked, no one was quite clear. Some believed that growing crops, trees, and shrubs changed the balance of moisture in the atmosphere. Others were convinced that railroads and telegraph lines acted as electrical conductors, "partially equalizing the electrical state of the earth and the air" and thus causing more rainfall. In fact, the science (or lack of it) behind Wilbur's idea didn't matter much. To many homesteaders, the notion that they could raise a great bounty of food and crops in a naturally dry landscape was just another modern miracle — one that was tied into a vision of settling the West as America's manifest destiny. The railroads were all too happy to promote this idea. The Santa Fe Railroad printed an official-looking map that showed the rain line moving west about eighteen miles a year with each new town tied to the railroad.

We now know, of course, that rain does not follow the plow. Thanks to their faith in bogus science and the people who promoted it, early homesteaders on the northern plains were doomed. Many of them arrived during an inordinately wet cycle. For eight years before 1887, the rainfall was bountiful in Colorado and western Kansas and Nebraska. Then the rain stopped, and the homesteaders watched their dreams dry up and blow away. Many abandoned their homesteads and moved farther west, to California and the Pacific Northwest. Others clung to the land, determined to ride it out and wait for the rain to return.

Instead of rain, however, they got rainmakers. "What was most difficult about the psychology of drought was the feeling of helplessness in the face of uncontrollable forces," the editor of the nation's leading medical journal wrote in 1925. "In times of stress, pain, or sorrow, science and truth give way momentarily to less rational approaches, and the sufferer is ready to leap at any cure or suggestion that may be offered to him for the alleviation of his travail, never stopping to inquire as to the motives of those who would heal him or as to the basis on which their claims may rest."

Of all the early-twentieth-century weather hackers I came across in my research, none offers a better cautionary tale for geoengineering than Charles Hatfield, who was, in his day, the most famous rainmaker in the world. In a profession rife with swindlers and con men, Hatfield was seen by some as the real thing—"the Henry Ford of rainmaking," one observer called him. In part this was because of Hatfield's sober, unflashy manner and his habit of wearing a nice suit and a fedora. But mostly it was because, unlike most rainmakers, Hatfield had actually pulled rain out of the sky. Or so it seemed to many in California, where Hatfield often plied his trade. Of course, there were plenty of skeptics, but just as many people believed that he understood something about how weather worked that others didn't.

During his career, which lasted from about 1910 to the late 1930s, Hatfield ranged all over the West, as far north as the Yukon and as far south as Guatemala. In Los Angeles, where he had several high-profile successes, it was never "raining"—it was "Hatfielding." Enterprising merchants sold "genuine Hatfield umbrellas" for $4.50 each. On the lecture circuit, Hatfield's talk, "How I Attract Moisture-Laden Atmosphere," routinely sold out. When he visited Alaska, three thousand people and a brass band met his boat. He talked openly about dissipating London fog and irrigating the Sahara. In Canada, members of Parliament—foreshadowing the debates we're sure to have over geoengineering in the years to come—worried that if they allowed Hatfield to try making rain in the northern territories, he might damage "the vast and delicate atmosphere of the universe." But it was in San Diego, "the City of Miracles," that Hatfield had his greatest adventure, during the winter of 1916.

Since its founding in 1769 as a base of operations for Spanish explorers and missionaries, San Diego had grown into a booming resort, a place blessed by a magnificent harbor, natural beauty, and eternal sunshine. On a visit to San Diego in 1872, Swiss naturalist Louis Agassiz, a towering figure in the history of climate science, foresaw a bright future for the city: "This is one of the most favored spots on the earth, and people will come to you from all quarters to live in your genial and healthful atmosphere." All along the beach, grand hotels reminiscent of Newport, Rhode Island, and Saratoga Springs, New York, sprang up—wooden white elephants with names such as the US Grant, the Lincoln, and the Hotel del Coronado (where Marilyn Monroe later frolicked with Jack Lemmon and Tony Curtis in *Some Like It Hot*). The Panama-California Exposition, which was set to open in Balboa Park in the early spring of 1915, would celebrate not just San Diego's role as the first port of call for the newly completed Panama Canal but also its coming of age as a major city on the West Coast and a potential rival of San Francisco and Los Angeles.

There was only one problem: in its natural state, Southern California is a desert. To bloom, it needs water. And the farther south you go, the drier it gets. For early explorers, this was not a big issue. But the natural cycle of rain was not enough to sustain the dreams of San Diego's civic leaders and real estate speculators, many of whom saw their future limited only by the number of gallons of water they could wring from the clouds.

Drought was not an abstraction in turn-of-the-century Southern California. Many residents had lived through a seven-year dry spell during the 1890s. They had seen apricot trees wither, wheat dry up, and strawberry fields die. The human toll was worse: farmers went bankrupt, lost their farms, and fell into depression. Some fled; others hung themselves. The city of Los Angeles lost half its population in a few years. Those who stayed learned a lesson they would never forget: the dream of the West as an earthly paradise was entirely dependent on a few inches of rain.

In some ways, San Diego was not as vulnerable as LA. The population was much smaller. It was a resort economy, not an agricultural one (although that was changing quickly in LA, thanks to the discovery of oil and the exploding popularity of motion pictures). But LA, unlike San Diego, had taken a big step toward solving its water problems with the construction of a 233-mile-long aqueduct from the water-rich Owens Valley. The aqueduct, an epic engineering project masterminded by William Mulholland, the superintendent of the Los Angeles Department of Water and Power, had changed the destiny of the city, breaking it free of nature's constraints and opening the door to a century of growth and development.

San Diego didn't have a William Mulholland, but it did have John D. Spreckels, heir to a sugar fortune, who developed railroads and shipping lines there. What Spreckels lacked in vision he made up for in cash. In the 1880s and 1890s, he built a series of reservoirs in the high, dry mountains, connecting them with a network of canals, rivers, and wooden troughs. Although he had conceived of this water project as a business venture, he eventually lost interest in the idea and sold the entire system to the city at a discount rate. It was an improvement over the old water system, which consisted of nothing more than private wells and a poorly engineered diversion of the San Diego River. For a while, it seemed that San Diego's water problems were solved. In Balboa Park, for example, a new fountain shot water 125 feet into the air — a gesture of extravagance that would have been unthinkable a few years before.

By 1915, however, it was clear that the city's water problems were far from over. Rainfall in the area had been about average over the previous few years, but the city's reservoirs were only a third full. Even worse, thanks to the city's rapid growth, demand for water was rising fast.

Among the members of the San Diego City Council, desperation was palpable. A water famine in the spring, just as attendance at the Panama-California Exposition was at its peak, would devastate the city. What could be done? There was no time to build more reservoirs and canals. Rationing water was unthinkable. There were no Wal-Marts stocked with bottled water from Maine or Tahiti. Desalinization plants had not yet been invented. And, of course, they could not squeeze rain from a cloudless sky. Or could they?

Charles M. Hatfield was born in Fort Scott, Kansas, in 1875. The Hatfields were a Quaker family, and young Charles grew up with the idea of "God being within"—a fact that may have added to his lifelong belief that he had the power to summon rain. His father was a property speculator who was drawn to California by good weather, rich soils, and the smell of money. When Hatfield was about ten years old, the family moved west and settled among the apricot and prune orchards near Melrose Boulevard in Hollywood. The movie business had yet to be invented, and Hollywood was still a bucolic farming community. Young Hatfield, however, was not much of a farmer. He and his younger brother Paul spent most of their free time dismantling and reassembling sewing machines. When he was in the ninth grade, he quit school and took a job as a salesman for the New Home Sewing Machine Company in Los Angeles. Tall, thin, handsome, and always dressed in a carefully pressed suit ("He was a dandy," his brother Paul said), Hatfield quickly became one of the company's most successful salesmen.

But Hatfield had bigger ambitions. In the mid-1890s, when he was about twenty years old, California was hit by the seven-year drought that spurred the development of Mulholland's aqueduct. Hatfield would later say that seeing the poverty and hardship the drought caused inspired him to undertake a "deep and serious study of the science of the atmosphere."

More likely, he sensed opportunity. While he was selling sewing machines in LA, he followed the adventures of the rainmakers in the local newspapers, especially the cloud bombers who claimed to suck water out of the sky with explosives. Hatfield thought they were frauds and had a hunch that rain could be coaxed from the sky in a gentler way—with chemicals. In this, he was not altogether original. "Fume men," as they were called, were a distinct subculture of rainmakers. Although they didn't offer the shock and awe of the boomers, they had an air of mystery and sorcery that was hard to resist.

Hatfield claimed that he found his inspiration in a boiling kettle. He believed if he released vapors from certain chemical combinations near the kettle, the steam from the kettle would move toward the chemicals. After what Hatfield described as a "deep and prolonged study of the cause of certain known phenomena in nature," he thought he had found a kind of meteorological magnetism. Unlike other rainmakers, Hatfield was careful not to overstate his powers. "I do not make rain," he told one inquiring reporter. "That would be an absurd claim. I merely attract the clouds and they do the rest." He was not a rainmaker but "a moisture accelerator."

Although Hatfield was deeply secretive about his methods, his basic technique was to mix up a witch's brew of chemicals—"a sort of meteorological mustard plaster," one observer called it—then allow it to age for a few days in metal containers. He often built a twenty-foot-high wooden tower to give himself both privacy and unimpeded access to the sky. He then poured the chemicals into a series of galvanized evaporation pans atop the tower. The heat of the sun did the rest, drawing the chemicals up in invisible fumes. Only a mild odor was given off, according to Hatfield, although one editor thought otherwise, writing that nearby farmers might be convinced that "a Limburger cheese factory has broken loose . . . These gases smelled so bad it rains in self-defense."

All these turn-of-the-century rainmakers, Hatfield included, were, at best, delusional. You can't make rain with bombs, electricity, or zinc fumes. Hatfield's main advantage was that he had studied rainfall patterns and averages, so he knew when it was time to set up his "precipitation plant" for the maximum likelihood of rain. In effect, he knew how to play his hand. This is not to suggest that he didn't believe in his own powers. By all accounts, he did. It's just that he was smart enough to leverage those powers with hard data.

Hatfield's first notable success came in 1904, when Los Angeles was suffering from a three-year dry spell. In late 1903, he approached the Los Angeles Chamber of Commerce and promised it at least eighteen inches of rain by late April in exchange for $1,000. Normal rainfall rarely exceeds eight to ten inches, he noted, adding that he would take no pay whatsoever should less than eighteen inches fall. Given the city's desperation, who could resist? The $1,000 was raised within three days, with a number of prominent businessmen contributing.

Hatfield went to work immediately. He got big play in the newspapers, a mix of awe and mockery. Angelenos laughed and, half-jokingly, begged the young engineer not to let it rain on the Tournament of Roses parade. But they could not dismiss him entirely. "The suggestion that a man of this class should be given serious consideration is apt to provoke a pitying smile," wrote one reporter for the *Los Angeles Examiner.* "But it is always to be remembered that the same pitying smile was wasted on . . . [others] who did that which seemed to the men of their time impossible. There is always the haunting fancy that he really may be able to produce rain at will." And if that was true, the writer speculated, what if Hatfield tapped into powers beyond comprehension? What if he "produced a storm monster that he [could not] control? He may be the Frankenstein of the air."

In March, the rain came. By June, twenty inches had fallen. Hatfield received his $1,000—and a boatload of publicity. "The people's prayers have been answered through my son," his mother told the *Los Angeles Times.* "For five years he struggled against prejudices. His determination is simply marvelous. Some divine power must aid him." When people asked how he did it, Hatfield demurred, insisting that his formula remain secret. "I do not fight nature," he said. "I woo her by natural means."

Lake Morena is sixty-five miles east of San Diego, in the high desert hills not far from the Mexican border. It's rattlesnake country, a rocky, remote place where the lake's deep blue water looks unnatural, like an alien substance, but is all the more beautiful because of it. When I visited the reservoir in the fall of 2008, the lake was at about 10 percent capacity—a little lower than it was in Hatfield's time. I could see the high-water mark on the rocks above the current water line, a reminder of the drought that was hitting California at the time. Rainfall in the state was only about 30 percent of average, the lowest in 114 years. A few months before my visit, Governor Arnold Schwarzenegger had declared a statewide drought alert. Nine counties, mostly in the Central Valley, were so dry that a drought emergency was declared.

When Lake Morena was completed in 1912, it was the top supplier of water to San Diego, the highest of six reservoirs that fed the city. Today its role is much less crucial. San Diego imports about 80 percent of its water from the Sierra Nevada to the north and the Colorado River to the south. In this highly engineered system, Lake Morena is little more than an emergency backup supply. But in 1916, there was no backup. If the lake went dry, so did the city.

A bronze marker set in stone marks the spot where Hatfield's precipitation plant stood in the winter of 1916. No other sign of his presence remains, but it is not hard to imagine the forty-year-old rainmaker up in his wooden tower, fedora tilted back on his head; blue eyes squinting up at the sky, looking for a wisp of a cloud on the horizon; hand reaching for the brass barometer that he wore on his belt, ready to register any drop in pressure that might signal an approaching storm.

Hatfield's encampment on the shores of the lake resembled a small, fenced-in military outpost. He and his brother Paul had built it themselves, on a slope beside the road leading to the dam. Within the encampment were two large, white canvas munitions tents—one for sleeping, the other full of chemicals, evaporation pans, and other equipment. Nearby was the wooden tower and a flagpole flying Old Glory. (The fact that he took the trouble to erect a flagpole tells you something about how seriously he took his work.) While Hatfield was up in the tower, cooking up the chemical stew that he believed would coax rain from the sky, Paul could often be seen pacing the perimeter fence, prepared to welcome visitors with a rifle.

Hatfield had his agent, Fred Binney, to thank for the job at Lake Morena. In 1912, after a winter of less-than-average rainfall, Binney, on behalf of the San Diego Wide Awake Improvement Club, had sent an urgent letter to the city council. "The Morena Dam reservoir is barely one-third full, and the city's growth hinges on an ample water supply," he wrote. "We think the council should consider hiring Charles M. Hatfield to make some rainfall." Binney proposed paying Hatfield $5,000 to produce an unprecedented five inches of rain between May and September. The council turned him down. In late 1915, Hatfield himself appeared before the city council. This time, with the Panama-California Exposition in full swing and the city brimming with tourists, the fear of a water famine loomed large. "I will fill the Morena reservoir to overflowing between now and next December 20, 1916, for the sum of $10,000, in default of which I ask no compensation," Hatfield proposed. After brief discussion, the council agreed by a four-to-one vote to accept Hatfield's proposition. However, no formal written contract was drawn up — an oversight that Hatfield would regret for the rest of his life.

Hatfield began operations promptly on January 1, 1916. For a few days, nothing happened. The dam operator, the only other inhabitant at the lake, reported seeing Hatfield in the tower day and night, his sleeves rolled up, working hard with the evaporation pans. At night, to keep the evaporation going, he'd heat the pans on a wood stove.

On January 5, a good rain fell, then tapered off. Then on January 10, it started to pour, and it didn't stop. Roofs leaked. Storm drains that had not been tested for years overflowed. On January 17, the San Diego River overflowed its banks, sending water pouring into the streets of the city, washing out railroad lines, flooding small towns in the valleys, and putting pressure on other reservoir dams beneath Lake Morena, especially the Lower Otay Dam. The hardest-hit area was Little Landers, a utopian community of a hundred or people living off the land. The rising waters destroyed their homes and washed away their soil. Two women drowned.

That morning in the *San Diego Union*, this headline ran above pictures of the devastation: "Is the Rainmaker at Work?" The paper reported, "The mysterious Hatfield, rainmaker, was said to be particularly active in the vicinity of Morena Sunday . . . While engaged in his experiments, Hatfield is not altogether sociable, but persons watching his work from a distance said he seemed to be on the job all hours of the day and considers the downpour due to his efforts. Incidentally, it was said that Hatfield is getting a good soaking."

Then, as quickly as it started, the rain stopped and the sun came out. Repair crews went to work on the roads and bridges and railroad lines, farmers rounded up their animals, and burial arrangements were made for the two women in Little Landers.

At Lake Morena, Hatfield was unaware of the chaos below. There were no phones, no BlackBerrys, no smoke signals, no way to communicate how much damage the rain had caused. Hatfield was completely focused on filling up the reservoir. He watched the water level rise each day, as the runoff flowed down the swollen creeks and into the lake. What power he must have felt, what a sense of connection with the sky and the clouds overhead! Hatfield had kept working through the rain, believing that his fumes would help keep the moisture flowing out of the clouds.

After a few days, as if on cue, another wall of dark clouds rolled in. The rain was even heavier than before. Hillsides, soaked from the previous week's rain, began to collapse. The river boiled up again, washing out dozens more roads and bridges.

By January 27, the dam at the Lower Otay Reservoir began to fail. For several hours, city workers shored it up with timber and steel, but it was too late. The dam collapsed, sending a surge of water forty feet high down through the valley, wiping out several communities and carrying a wall of debris—houses, animals, bodies—all the way out to the Pacific Ocean. The next day, after the rain stopped, the coroner's office estimated that fifty people had been killed. (That number was later revised down to twenty.) Property damage exceeded $3.5 million. And amid the soggy wreckage, people questioned Hatfield's role in triggering this deluge. Was the rainmaker to blame? Had he squeezed the clouds too hard? If nothing else, Hatfield was an easy scapegoat. There was talk of organizing a party to travel up to Lake Morena and lynch him.

Hatfield and his brother, still unaware of the disaster below, hung around for three days, dismantling the tower, packing up their equipment, and preparing for their triumphant return to San Diego. Hatfield even raked the ground where the tower had stood, just to make it all neat and tidy. When he looked out at the lake, he must have felt like Superman: it was full to the brim with fifteen billion gallons of water.

Hatfield learned about the disaster from the Morena Dam superintendent. Apparently, someone had found a way to let him know and perhaps to warn him that Hatfield's life could be in danger. Hatfield and his brother quickly disappeared on foot down the trail below the dam, heading off in the general direction of San Diego. Both men were armed.

There is no record of Hatfield's reaction upon seeing the destruction — the bloated bodies of horses and cows, the washed-out roads and bridges, the wrecked homes. But his behavior suggests that he knew he was in a dangerous situation. Perhaps he was thinking about the fate of Australian rainmaker Frank Melbourne, who was found dead under mysterious circumstances in a scruffy hotel room in Denver in 1894, his body identifiable only by the monogram on his clothes. Had Melbourne finally hustled the wrong man? When Hatfield and his brother finished their two-day, sixty-mile trek back to San Diego, they lay low for a few days at the home of his agent before the local press tracked him down.

It may or may not be a sign of Hatfield's delusional character that when he emerged from hiding, according to one reporter, he had the demeanor "of the proverbial conquering hero, home from the fray and awaiting the laurel wreath." Hatfield said that he had lived up to his promise to fill Lake Morena and would now formally lay claim to his $10,000 before the city council. He'd filled up the reservoir, hadn't he?

Hatfield's predicament raised a question that's still relevant today: when it comes to climate, how do we know what is "natural"? After Hurricane Katrina wiped out New Orleans and other parts of the Gulf Coast in 2005, climate scientists and skeptics debated whether global warming was in any way responsible for the storm. Just a month before it hit, Kerry Emanuel, a hurricane expert at MIT, had (coincidentally) published a high-profile scientific paper describing how warmer waters in the Gulf of Mexico can rev up the heat engine that creates hurricanes, increasing their intensity. Some global warming activists were quick to seize on this as evidence that Katrina's power and devastation were, at least in part, a result of rising levels of CO_2 in the atmosphere. Advertisements for Al Gore's *An Inconvenient Truth* even used an image of a hurricane swirling out of a smokestack, implying a clear link. This is not just a matter for scientific debate. If rising CO_2 emissions can be demonstrated to cause more intense hurricanes, then big oil and coal companies could be held responsible for the damages caused by these storms. In effect, they would be in the same position as Charles Hatfield—individual actors accused of nudging the climate in a way that damages us all.

The problem is determining cause and effect. Or, to put it another way, where does nature end and human influence begin? Higher CO_2 levels may indeed pump up the thermodynamic engine that creates hurricanes, making it more likely that they will form and adding intensity when they do, but that does not mean that any *particular* storm can be attributed to the fact that we've been on a two-hundred-year fossil fuel binge.

It becomes even more complicated when you consider directly manipulating the earth's climate to offset global warming. Once we admit to trying to steer the climate one way or the other—by creating artificial volcanoes to pump particles into the stratosphere or manufacturing clouds or using some other method—then whoever undertakes this effort steps, in effect, into the shoes of Charles Hatfield. If the man-made clouds have the desired effect, the geoengineer is a hero. If they do not, or if, for whatever reason, the region is hit by unusual weather conditions, it will be all too easy to hold that person responsible for whatever weirdness happens. "We spend a lot of time arguing about the weather now," Richard Alley, a leading paleoclimatologist at Pennsylvania State University, told me. "Imagine what it will be like if I can blame somebody every time my tomatoes don't ripen on schedule."

This does not mean we're stuck on a merry-go-round of random events. The more scientists learn about how the climate system works, the more certain they can be about cause and effect. Consider what has happened in other fields. Thanks to rigorous epidemiological studies, we know that smoking causes cancer, even if scientists don't know the exact triggering mechanism. Ditto for the health impacts of air pollution. In fact, it's not hard to imagine a day when climate models become good enough that scientists can actually make reasonable guesses about which weather events are "natural" and which are the result of human interference with the climate system. But that day has not yet arrived.

And it certainly was not there in 1916, when Hatfield appeared before the San Diego City Council to collect his $10,000 for filling Lake Morena. Hatfield's claims were complicated by the fact that during most of January, it had been raining all up and down the coast of California, even as far north as Oregon—far beyond what even Hatfield's most credulous supporters could ascribe to his influence.

Hatfield, not surprisingly, didn't claim to have brought rain to the entire West Coast—only to have helped maximize the precipitation in the region around Lake Morena. (According to the city water department, twenty-eight inches of rain fell at Lake Morena in January, far above the monthly average.) Lawyers for the city peppered Hatfield with questions: How much time had he spent at

Lake Morena? If he was going to claim credit for the rain, would he also accept liability for the damages? How about the deaths?

The city council's lawyers spent a few months wrangling over this — they did have a verbal contract with Hatfield, after all — before ultimately deciding that because Hatfield could not "prove" that his actions had caused the rain that filled the lake, the city owed him nothing. From a scientific point of view, it was a sensible decision. We know now that the flooding of 1916 was likely the result of an El Niño weather system that was parked out in the Pacific. This type of system typically drives heavy storms onto the West Coast. From a legal point of view, however, the decision was more complex. Just as Hatfield could not prove that he had filled the reservoir, the city could not prove that he had not. Hatfield, bitterly disappointed and feeling betrayed, pursued his case against the city in court for years, without success.

After the San Diego flood, Hatfield continued to romance the clouds in California and around the West, as well as in Europe and Honduras, where he tried to stop a forest fire (he failed). Eventually, however, rainmakers fell out of fashion, done in by the increasingly vocal skepticism of the scientific establishment, as well as the increasing sophistication of public works projects. If you can get water reliably from your tap, who needs a rainmaker? David Starr Jordan, the former president of Stanford University, published a scathing essay about rainmaking in *Science* in 1925, comparing it to "crystal-gazing, clairvoyance, and kleptomania." By 1956, when Hollywood made *The Rainmaker*, starring Katharine Hepburn and Burt Lancaster and based partly on Hatfield's life, rainmaking could be safely portrayed as something akin to faith healing or voodoo. Hatfield himself had long since faded into obscurity, dying at his home in Southern California in 1958.

It's easy to write Hatfield off as a crank, a relic of an earlier, more primitive civilization. And in many ways, he certainly was. But geoengineering is not rainmaking. Scientists who talk seriously about manipulating the earth's climate aren't con men (not that there aren't a few crackpots loitering in the shadows, talking about dumping tons of Special K into the ocean). They are, in fact, rational, ethical, deeply serious people who are concerned about the trouble we've gotten ourselves into and are looking for any tool that might help us get out of it.

Every era has its own brand of arrogance, however. We look back fifty or a hundred years, and we think that simply because we can check the weather and stock prices on our iPhones, we are immune to the peculiar mix of arrogance and innocence that led people to put their last dollar in a hat to hire a man who told them with great certainty and faith that he could call rain down from the sky. We laugh, thinking how naive they were, how simple-minded.

But that may only be because we haven't known desperation. A few months before my visit to Lake Morena, I happened to spend the weekend in Atlanta. The region was in the midst of a fearful drought—water levels at city reservoirs were dangerously low, watering lawns and washing cars were prohibited, real estate developers were in a panic, and the whole city seemed to be ready to shut down if it didn't rain soon.

I checked into my hotel and flipped on the TV. The lead story on the local news was Georgia governor Sonny Perdue, standing red-faced and worried in front a group of farmers in Macon, leading a public prayer for rain.

FOUR

Big Science

YOU COULD ARGUE that the modern era of geoengineering began on July 6, 1962, when technicians armed a 104-kiloton nuclear
device at the U.S. Atomic Energy Commission's Nevada Test Site,
about sixty-five miles north of Las Vegas. Unlike most previous
"shots"—test site lingo for nuclear explosions—this bomb would
not be detonated from a tower several hundred feet above the desert or while dangling aloft from a balloon. In fact, it was buried
635 feet below the desert floor. Nor was the purpose of this device,
which packed five times the explosive power of the bomb dropped
on Hiroshima, to test radiation fallout patterns or bomb triggers.
It was more elemental than that. The purpose of this experiment,
code-named Project Sedan, was to see how large a hole the bomb
would blast in the ground—that is, to see how well a nuclear bomb
worked as a high-tech earthmover, one that eventually could be deployed to move mountains, reroute rivers, blast canals, and generally sculpt the planet to make it a more comfortable place for human beings to live.

A few miles away from the detonation site, a group of scientists and engineers crouched in a bunker to witness the explosion.

Among them was Edward Teller, the Hungarian-born nuclear physicist and godfather of the hydrogen bomb. Teller, who was one of the models for Dr. Strangelove in Stanley Kubrick's black comedy about the Cold War, had eyebrows like the wings of a B-52, a prosthetic leg (due to a streetcar accident when he was a child), and a deep belief that American freedom depended on ever more powerful and sophisticated weaponry. As one science historian put it, "[Teller] never saw a new type of nuclear weapon that he didn't like, never liked any argument against building them bigger, smaller, more numerous, or more easily deliverable, never regretted anything about his career as the greatest of nuclear weaponeers except for the obstacles put in his way by those lacking in patriotic fervor or common sense."

But Project Sedan represented a turning point for Teller. It was the first test of a nuclear device whose explicit purpose was to demonstrate to the world that nukes were not just instruments of terror; they were tools of progress, too. Sedan was a key part of Project Plowshare, the U.S. Atomic Energy Commission's two-decade-long program designed to investigate and promote the development of "peaceful" and "constructive" uses of nuclear explosives. If the Sedan shot went well, Teller believed, humankind would be one step closer to one of his most luminous dreams—harnessing the limitless power of "clean" nuclear devices to reshape the planet, leading to a golden age of human welfare and prosperity. As Teller wrote, "We will change the earth's surface to suit us."

It was a statement of profound hubris, of course, but it also spoke to a central truth of the nuclear age: in humankind's long battle to subdue nature—one that many scientists would argue began with the birth of agriculture—the detonation of the atomic bomb represented a kind of victory. For the first time, humans had acquired a power that was greater than any thunderstorm, hurricane, tornado, or earthquake. Why be humble about it? We had access not just to the inner secrets of the atom, but also to other mind-boggling technologies born during the Cold War, including satellites and computers. Unlike in the days of the rainmakers, the constraints on how much human beings could do to shape the world were no longer technological — they were social, cultural, and, most of all, moral. Now that we could change the earth to suit us, would we? Should we? With Project Sedan, we finally had a tool that would let us reimagine our world on a grand scale. And for the first time, we were ready to put it to use.

Teller and his colleagues in the bunker at the Nevada Test Site pulled the trigger on Sedan at 10:00 A.M. In the archival films of the blast, you can see what happens next: a bubble of earth eight hundred feet across and three hundred feet high bulges above ground zero, pushing up from below. An instant later, one historian wrote, "incandescent gases burst through, shooting desert alluvium 2,000 feet out in a great hemispheric fountain of rock and dirt before the heavier material showered back to earth, partially filling the crater. A base surge rolled out radially, expanding to a distance of two and a half miles."

One eyewitness remembers it differently: "There was a boom, and a white flash, and then the biggest dust cloud you have ever seen. I was afraid we had cracked the earth wide open."

In the story of geoengineering, no one looms larger than Edward Teller. The weapons labs that he presided over were hothouses of scientific research, and many of the same questions that concerned bomb makers, such as the fallout patterns for radioactive material after an atomic blast, led to basic research that is directly relevant to climate scientists today. But Teller was influential not just because of his science but also because he so clearly embodied the Cold War faith in technology and engineering as the solution to every human problem. Whether it was protecting the White House from a nuclear attack or preventing civilization from being frozen out by another ice age, Teller believed it was all fixable, given the right tools and enough ingenuity. This is, of course, the fundamental ideology underlying the entire notion of geoengineering — the solution to global warming is not only to reduce emissions and change our lives but also to build better tools and become better climate mechanics.

But Teller is also important because he represented something entirely new in the American cultural landscape: the scientist as superhero. Teller and his archrival, J. Robert Oppenheimer, director of the Manhattan Project at Los Alamos, "presided over the transformation of theoretical knowledge into practical application, of chalkboard equations into explosive reality," wrote Steven Shapin, professor of science history at Harvard University. "Hiroshima and Nagasaki profoundly altered public perception both of what scientific knowledge was and what scientists were capable of delivering." Scientists such as Oppenheimer and Teller, Shapin wrote, held the keys to the magic kingdom: "With the advent of the Bomb, almost all scientists—not just nuclear physicists—began to appear as sources of power, and the extent to which American science fed off that legacy during the Cold War cannot be overemphasized." In a sense, today's geoengineers are all the mutant offspring of Teller and the Cold War bomb builders.

The big hole that Teller blasted in the Nevada desert back in 1962 is still there. In fact, it's one of the main attractions at the Nevada Test Site, which is open to a limited number of visitors one day each month. The test site is a ghostly place today, a 1,350-square-mile plot of desert near Las Vegas. Yet between 1951 and 1992, when a global ban on nuclear testing was passed, more than nine hundred nuclear devices were detonated here. Many of the glorious, frightening, iconic images of mushroom clouds rising in the sky were taken here. It was the Hollywood back lot for the Cold War.

On the day I visited the test site, the tour was awash with patriotic nostalgia. We left early that morning from the Atomic Testing Museum in downtown Las Vegas. In keeping with Vegas faux architecture, the museum is designed to look like an underground bunker. About thirty of us piled into a white, unmarked bus for the hourlong drive out to the test site. The other folks on the bus were mostly in their sixties and seventies, people who remembered the nuclear era as a glorious time in America, when the lines between good and evil were clearly drawn and America was a rising power.

We drove up Highway 93 into the desert. At the front of the bus, the tour guide, who introduced himself as Ernie Williams, ran through the rules for the tour: no cameras, no phones, no recording devices. Williams wore a hearing aid in each ear and spoke with military formality, often prefacing his remarks with "Ladies and gentlemen . . ." He was seventy-eight years old, he told us proudly, and had spent nearly fifty years at the test site, working as everything from bomb technician to comptroller. He had even played penny-ante poker with Teller. ("Mr. Teller was the finest gentleman you'll ever want to meet," he told us. "He treated everyone the same, whether you were a scientist or a janitor.") Over the years, Williams had witnessed eighty nuclear shots. He described what the shock wave from a nuclear blast felt like as it passed over your head at 550 miles per hour ("a big whoosh of air"). He handed around plastic-covered photos of towering mushroom clouds. They were smudged with fingerprints and seemed vaguely pornographic. "It was a great experience," he said of his years at the test site. "I'd do it all over again tomorrow." He assured us that despite five decades of exposure to radiation —"I'm pushing sixteen roentgens," he said, using the technical term for radiation units—he was in good health. "Ladies and gentlemen, radiation is like electricity," he said happily. "If you give it due respect, it won't give you any trouble."

In the town of Mercury, we turned off the highway and entered the test site. A well-armed man from a private security firm boarded the bus and checked everyone's ID, then we motored past the sad little town, full of Quonset huts and abandoned hotels where workers once slept fourteen men to a room. As we headed into the vast openness of Yucca Flat, the main testing ground, tumbleweeds skittered by, and the sky was empty of birds. At News Nob, a vista point from which Walter Cronkite once broadcast stories of nuclear tests to an awestruck nation, rows of sun-warped wooden benches sat empty, like relics of an earlier civilization. Steel fences ringed sub-

sidence holes, marking the spots where the earth above underground tests had collapsed. A wooden house, built in the 1950s to
examine the effects of a nuclear bomb on a typical American home,
now had the sorry, forgotten look of an old homesteader's cabin.

Finally, at the northern edge of Yucca Flat, in what is known as
Area 10 (Area 51, the top-secret military base, is just over the hill),
we turned off the main route onto a small dirt road. On both sides,
stark yellow and black signs greeted us:

DANGER

RADIATION AREA

NO DIGGING

A mountain of dirt came into view — the edge of the Sedan Crater.
People stood up in their seats before the bus came to a stop.

We filed out into the glaring sunshine, heading over to a well-built
metal viewing platform. I paused to read a bronze plaque, surprised
to find that the crater is listed on the National Register of Historic
Places. Everyone crowded around, staring down into the pit, which,
another sign told us, is 1,280 feet across and 320 feet deep. "This is
the pride and joy of the test site," Williams announced. "When it
blew, the dust cloud was five miles wide." The blast registered 4.75
on the Richter scale — about the same as a modest-size earthquake.
"The fireball was eleven thousand feet high!"

Williams talked briefly about Project Plowshare. As he explained
it, the project brought new pride to the workers at the test site: they
weren't weaponeers anymore; they were civil engineers. "We were
asked, 'Can we build a new Panama Canal?' We said, 'Yeah, but we
need to do a little homework first.'"

The Sedan Crater was the homework. I looked down the steeply
slanting earth walls of the crater. Rusty old pipes lay at the bottom;
bushes grew in the sandy soil of the walls. There was something
spectacular about this giant hole in the ground — a hole that was
dug, Williams pointed out, "in less than two seconds forty-five years

ago." When the device was triggered, he was in a tower six miles away. As he described the moment of detonation, his voice swelled with joy and nostalgia.

Redesigning the earth—with bombs or bulldozers or whatever tools were at hand—was not just a personal fetish of Edward Teller's. It was a staple of Cold War culture. Scientists in the old Soviet Union were particularly enamored with it. In 1949, the year the Soviets developed their atomic bomb, the Kremlin announced that it was employing nuclear explosions for economic development projects: "We are razing mountains. We are irrigating the deserts. We are cutting through the jungle and the tundra." It was all propaganda, of course, designed to drive men like Teller crazy. Still, even if they were not using nukes as shovels, the thought was there.

In 1960, two Soviet scientists published *Man Versus Climate*, a book that articulates as well as any the Stalinist vision of how to build a better world. As you might expect, the Soviets were very interested in thawing the cold north: "The Arctic ice is a great disadvantage, as are the permanently frozen soil (permafrost), dust storms, dry winds, water shortages in the deserts, etc." Ergo, "if we want to improve our planet and make it more suitable for life, we must alter its climate." They wrote about redirecting the Congo River and turning the Sahara into lush cropland; melting Arctic ice by covering it with ash, causing it to absorb more sunlight; dumping alcohol into the Gulf Stream to reduce evaporation rates and change the flow pattern; even building dams across the Dardanelles and the Strait of Gibraltar so that the Mediterranean would dry up, promising to recover "enormous areas of highly fertile land for man's use, to yield a fantastic amount of cheap electricity and, finally, to unite the continents of Europe, Asia, and Africa."

In contrast to the Soviet Union's outlandish ideas, Project Plowshare seemed downright tame. It was launched in 1957 at the Lawrence Livermore National Laboratory, Teller's scientific stomping grounds. The project name came from a physicist at the lab who, upon hearing about the idea of using nuclear weapons for peaceful purposes, remarked dryly, "So you want to beat your old atomic bombs into plowshares." The program was what today we might call a "rebranding" campaign for nukes. "It connected our work with reality," Teller explained.

But in Teller's view, Project Plowshare was also very much about finding justifications for continued nuclear testing, which was stirring up controversy around the world. In 1958, Linus Pauling, the Nobel Prize–winning chemist, presented a petition to the United Nations, signed by more than nine thousand scientists, calling for the end of nuclear testing. The petition declared that "each added amount of radiation causes damage to the health of human beings all over the world and causes damage to the pool of human germ plasm such as to lead to an increase in the number of seriously defective children that will be born in future generations."

Teller, of course, didn't buy any of this. "Unfortunately," he lamented, "much of the discussion about continued nuclear experimentation has been carried out in a most emotional and confused manner." For the average American, Teller believed, any increased risk of cancer and leukemia from nuclear testing was "minimal." Those who claimed otherwise were simply standing in the way of progress. (Teller, we now know, was wrong. According to a 1998 study authorized by the U.S. Congress, atmospheric testing between 1951 and 1962 was the likely cause of eighty thousand cancer cases in the United States, of which fifteen thousand were fatal.)

In Teller's view, once you got beyond the handwringing, the case for using nukes as excavators was pretty straightforward. "If anyone wants a hole in the ground, nuclear explosives can make big holes," he wrote in *The Legacy of Hiroshima*, his book-length argument for continued nuclear weapons development published in 1962. "The ability of nuclear explosions to move vast quantities of earth and rock—and to move them cheaply—opens a new and important discipline: geographical engineering."

To Teller, nature was an abstract space waiting to be engineered. There was no limit to what nukes could do. He dreamed about blasting new harbors in Africa, South America, and Australia. He thought that Panama could use a second canal. And maybe it would be useful to have a canal across the Isthmus of Kra, on the Malay Peninsula. Teller calculated that such a canal would cost a billion dollars to blast, but it would cut a thousand miles off the sea route between Japan and India—a great boon to the shipping industry. And how about improving the supply of fresh water in the world? Teller thought nukes could be employed to dig big craters to use as catch basins in the western deserts of Australia. He speculated about using nukes in Alaska to close the canyon through which the Yukon River flows and to make it flow in the opposite direction, where dams could then be built for electric power generation. And if Alaskans got tired of all the snow and cold, nukes could be used to, as Teller put it, "simply blast the ice pack, greatly roughing its surface and increasing its absorption of solar radiation."

Teller was a kind of technological Picasso, compulsively sketching out new ideas, re-creating the world according to his fancy. In the future, he thought, we might set off a ten-megaton explosion two or three miles underground, pour salt water into it, and then harvest the steam—voilà, a thermonuclear desalinization plant. Or ignite explosions to heat the Canadian tar sands, allowing the crude petroleum to be pumped to the surface in a liquid state. Or use nukes to shatter rock that holds oil shale, pushing out the hydrocarbons by superheating the rock. Even scientific research could benefit from nukes. If we set off a two-megaton bomb under, say, Antarctica, we could track the shock waves and learn something about the geophysics of the region.

His most enterprising proposal was to use nukes to manufacture diamonds. "High pressures near nuclear explosion could be used to compress some pure carbon until its atoms arrange themselves into this unique substance," Teller wrote. "With proper arrangement of materials underground, diamonds could be mass-produced."

As for radioactive fallout from all this geographical engineering, he had an answer for that, too. Because these nuclear explosions would be set off underground, most of the radioactivity would be trapped beneath the surface. According to Teller's calculations, no more than 15 percent could escape as gaseous radioactivity. "If we take the necessary precautions," he wrote, "we can be sure that no person will be exposed to radiation effects greater than everyone receives from natural sources."

Despite all this, Teller acknowledged that using nuclear explosives to shape the earth was not an easy sell. "To be an optimist requires courage," he wrote. "But to be an optimist in the nuclear age demands even more: it demands imagination."

As often happened at America's nuclear weapons labs, projects begat projects. Project Plowshare led directly to Project Chariot, the first attempt at a real-world test of nuclear earthmoving.

In order to win the argument that nuclear bombs could make the world a better place, Teller needed to prove his case. He decided that the simplest way would be to dig a harbor somewhere. What is a harbor but a deep hole near the shore? Excavation need not be exact. Best of all, we could easily argue that we were "improving" the landscape: harbors bring ships, commerce, money, political power. Even fish like deep harbors.

The Project Chariot team first targeted Arica, Chile, a remote spot on the west coast of South America with about thirty thousand people, for this test. They liked it because the population density was low, the geology and meteorological conditions were suitable (the fallout could be safely blown out to sea), and they thought the Chilean government would welcome the chance to improve trade routes in the region (not that they ever bothered to ask). But after a little thought, detonating a one-hundred-kiloton nuclear device off the coast of a foreign country didn't seem like such a good idea after all, and the proposal was scuttled.

The focus shifted back to the United States, with Alaska quickly falling into the cross hairs. It was still far enough away—culturally, politically, and geographically—but there would be no messy international ramifications to blasting a big hole on the coast. (*Hey, it's our country—we can blow it up if we want to!*) The planners

chose a spot on the northwestern coast, near the mouth of Ogo-
toruk Creek on Cape Thompson. When Teller flew over the area
with a bush pilot, it seemed to be the perfect canvas for his nuclear
doodlings: from five thousand feet, it looked like a vast expanse of
nothingness on a long spit of land about a hundred miles above the
Arctic Circle. Who could object to planting a few bombs there?

So Teller and his team from Livermore flew up to Alaska to court
local politicians and editorial writers. Spreading their atomic fairy
dust, they invited the Alaskans to hitch their wagons to the great engine of progress. Teller relished going over the technical details of
the new harbor, which, if all went according to plan, would be one
hundred to three hundred feet deep, four hundred yards wide, and
two thousand feet long. The excavation would take less than two
seconds, thanks to his plan to detonate five simultaneous explosives—three 150-kiloton devices and two 1.3-megaton ones. This
was 160 times the explosive power dropped on Hiroshima in August
1945 and about equal to 40 percent of all the explosive energy used
in World War II. Teller was so supremely confident of his technical
prowess that he joked with his colleagues that if it would make Alaskans happy, he could even carve the harbor in the shape of a polar
bear.

The truth was, Teller and his team were interested in blasting the
harbor because it would give them information on the cratering
capability of nuclear devices, which in turn would offer a new ra-
tionale for continued testing—and, by extension, continued bomb
building. But he pitched it to the Alaskan audience as a way to boost
economic development in the region, especially commercial fishing. The real payoff would come, Teller argued, when a railroad was
built and they began to export "the largest deposits of [the] highest
quality proven coal deposits in Alaska," which he said were adjacent
to Cape Thompson. In a statement that demonstrates Teller's skill
at framing the debate, he pointed out that Japan, which has no coal
of its own, currently imported coal from Pennsylvania. If the Japanese could get cheaper, more abundant coal from Alaska, he rea-

soned, they "might become the first beneficiaries of atomic explosions as they have been the first victims."

It sounded good to many Alaskans. "What objections could there possibly be to this large-scale atomic harbor-blasting project?" asked an editorial in the *Fairbanks Daily News-Miner*. "Scientists who have studied the entire matter carefully have given assurances that the project has been so carefully planned that blasts will have a minimum of fallout." And in any case, "the project is located in the wilderness, far away from any human habitation." In the end, the editorial concluded, "this vital project will result in incalculable benefits to all mankind."

But before Teller could get his bombs delivered and deployed, a few Alaskans started asking questions, especially about the political impact of radiation fallout. The border with the Soviet Union was only 180 miles across the Bering Strait. What if an unexpected wind shift sent radiation drifting over the United States' Cold War enemy? A few conservation types (there weren't many in Alaska at the time) pointed out that Cape Thompson itself might be wilderness, but it was only about thirty miles away from Point Hope, a village of about three hundred Inupiats. Had anyone asked them what they thought of this lovely new harbor? The Inupiats had been hunting and fishing around Cape Thompson for about five thousand years. In fact, Point Hope was one of the most successful communities in the Arctic. The Inupiats hunted caribou, whales, and walruses at Cape Thompson; they fished in the bay; they collected eggs from the sea birds. Even after meeting with the Livermore scientists and watching an animated movie of a nuclear device excavating a harbor, the Inupiats remained unconvinced that the detonation of a series of nuclear bombs in the area would improve their lives.

Teller did not lose much sleep over the fate of the Inupiats. In 1958, when one Alaskan economist pointed out that the Inupiats depended on sea mammals and caribou in the region for their survival, Teller responded, "They're going to have to change their way of life."

"What are they going to do?" the economist asked.

"Well," Teller said, "when we have the harbor, we can create coal mines in the Arctic, and they can become coal miners."

Despite Teller's best efforts, opposition to the project gained momentum. Within a year or so, it became clear that this "early and obvious demonstration," as Teller had once put it, would take a lot of time to develop — far more than the original eighteen months he had envisioned. The project team downsized the blast several times, hoping it would help dispel fears about fallout.

But it was a young geographer named Don Foote who finally killed off Teller's Alaskan dream for good. After flunking his Ph.D. exam at McGill University in Montreal in 1959, Foote decided to take some time off and do some fieldwork. He ended up getting hired by the Alaska Resource Development Board to carry out "human geographical studies" in the Cape Thompson area. In reality, the project was intended to be a rubber-stamp government study concluding that there was nothing much to lose by setting off a series of nukes in the region.

But Foote took the study seriously. Instead of phoning it in, he lived for more than a year among the Inupiats in a cardboard-walled hut. He cataloged their whalebone houses; learned about their hunting methods, their history, and their social structure; and came to understand the nature of the world around them. Far from a frozen wasteland, Foote determined, Cape Thompson was rich in a variety of flora and fauna. The cliffs along the cape were one of the principal nesting grounds of a huge variety of sea birds, including puffins, murres, gulls, and kittiwakes. Foote cataloged three hundred species of plants and twenty-one species of mammals, from collared lemmings to grizzly bears. And he discovered that, despite Teller's assertions to the contrary, the wind at Cape Thompson was unpredictable. Sometimes it blew offshore — straight toward the Soviet Union.

This was not the kind of information that Teller and the other Project Chariot team members were eager to hear. They tried to marginalize Foote, burying his studies and omitting environmental information from official reports. Foote was outraged. "My belief in you and the program has been shattered," he wrote to one of the project leaders. "To a great extent this was the effect of my politically immature mind awakened to the modern interplay of politics and science. It was caused by my realization that when science becomes directly involved with politics, the concept of scientific truth becomes modified, warped, and totally abandoned."

Foote got his revenge. In 1961, newspaper reports of the controversy over Project Chariot piqued the interest of Paul Brooks, editor in chief of Houghton Mifflin and a well-known writer interested in conservation issues. (This was a year before the publication of Rachel Carson's *Silent Spring*, which is often cited as the wellspring of the modern environmental movement.) Foote, working with his brother Joe, agreed to feed information about the project to Brooks. Brooks and Joe Foote coauthored an article in *Harper's Magazine* that drew international attention to Teller's plan and ultimately hastened its doom.

"Our ability to alter the earth we live on is already appalling," Brooks and Foote wrote, vividly foreshadowing today's debate about climate engineering. "Few of us are in a position to judge the ultimate scientific value of an experiment like the Chariot explosion. But it *is* up to us to know what is going on in a far corner of the United States. And to realize that another scale of values is also involved: not the precise relations between depth of burst and crater characteristics, but the precise relations between unlimited power and the awesome responsibility that goes with its use."

By the spring of 1962, it was well-known both within the Atomic Energy Commission (AEC) and the Lawrence Livermore Lab that Project Chariot was dead. But to simply cancel it would have been a public relations disaster for Teller's dream of finding a peaceful use for nukes, an acknowledgment that even a demonstration project in the wilds of Alaska was too fraught with controversy and complexity to succeed. So, instead, project scientists hurriedly assembled a new plan. Rather than blasting a hole in Alaska, they would blast one at the Nevada Test Site, where nobody could stop them, and the drama of the explosion, if nothing else, would distract everyone from Project Chariot's failure. Thus was born Project Sedan, the largest nuclear test in North America up to that time.

Within weeks, the 104-kiloton nuclear device was ready to go. There was no hesitation, no counting of birds, no worries about migrating caribou, no fretting about fallout over the Soviet Union. Technicians drilled a deep hole, the bomb was lowered into it, and, after a pause to clear the area, the bomb was detonated. The AEC immediately deemed it a success. Within an hour, the agency estimated that "95 percent of the radioactivity produced by the detonation was trapped in the ground." Of the remaining 5 percent, the AEC hastily concluded, "most" landed "close to the crater and within the test site."

They were wrong.

Later studies showed that once Sedan's radiation was aloft, the wind caught it and carried it north. At 2:45 P.M. on the day of the blast, the AEC had to backtrack on its earlier claims, announcing to the press that the cloud had crossed State Route 25, the main road near the test site. The Nevada State Police closed the road, and a road maintenance station was evacuated, as were people from nearby ranches. The AEC advised others to remain indoors while the cloud passed. The dust was so thick around Ely, Nevada, two hundred miles away, that the streetlights turned on at four in the afternoon. Five days later, the dust passed over Utah, Colorado, Wyoming, Nebraska, and the Dakotas. Two years later, AEC chairman Glenn Seaborg admitted that Sedan fallout had probably crossed into Canada. It took twenty-one years before an AEC report acknowledged that the explosion had "deposited nearly 5 times as much fallout on and near the test site as had been predicted."

So much for "clean" nuclear earthmoving. One historian calculated that a shot the size of Sedan, fired in Alaska, could have dropped radioactive fallout over the entire length of the North Slope or penetrated a thousand miles into Siberia, depending on the wind and differences in the geological character of the blasted rock. The Chariot shot, at its smallest configuration, would have been nearly three times as powerful as Sedan. At its largest configuration, it would have been twenty-four times larger.

Incredibly, Project Plowshare lived on for another decade. In 1964, the U.S. Congress authorized a five-year, $18 million study to explore the possibilities of blasting a new canal through Central America. In Mississippi, Plowshare scientists proposed using nuclear explosives to make a twenty-five-mile canal cut to connect the Tennessee and Tombigbee rivers. In California, they proposed using nukes to help straighten out Route 66. But even within the bomb factories, resistance grew. In 1965, John Gofman, head of Livermore's new biomedical division, informed the Livermore leadership that he believed "building a canal with hydrogen bombs would be biological insanity."

Finally, in 1970, after investing more than $700 million over two decades, Congress killed the nuclear excavation program, and Project Plowshare died a quiet death a few years later. "The story of Project Plowshare," wrote one Cold War political analyst, "is one of sustained and futile scientific and engineering optimism in the face of a world becoming increasingly intolerant of such ventures . . . The termination of Plowshare was the reluctant admission that a nuclear utopia was not imminent."

The end of Project Plowshare was not the end of Teller, who still had more than two decades of work ahead of him. In the 1980s, he became the central architect and advocate of the "Star Wars" missile defense system. His right-wing Republicanism and, as one critic put it, "fear and loathing of all things Russian" made him an ideal scientific sidekick to President Ronald Reagan. But Star Wars turned out to be a multibillion-dollar scientific boondoggle, yet another fevered technological dream that never quite worked out the way Teller imagined. After the Berlin Wall fell in 1989 and the Soviet

Union imploded, Teller was left without an enemy to shake his fist at, and he spent the final years of his life in his office at Livermore, a gimpy, half-blind old man, shunned by many of his scientific colleagues and increasingly bitter and reclusive.

But he still liked to think big. One problem that interested him in his final years was what would happen to the human race if an ice age returned. At the time, the long-term implications of global warming had not yet sunk in, especially among physicists such as Teller, and many scientists believed that climate cycles would inevitably return the planet to a deep freeze. When that happened, how would the human race survive?

Teller had only begun work on that problem when he died in 2003, at the age of ninety-five. At his memorial service at the lab, his colleagues talked about his "unapologetic optimism." "He believed that technology could improve most aspects of the human condition — perhaps even drastically so," one physicist said. "Very notably, he was convinced that all people — particularly his friends and most especially his fellow Hungarians — could rise above their human frailties somewhat and perform more like angels in the future than they had in the past."

It sounds odd to talk about Edward Teller and human frailties in the same breath, but there is a larger point here. Teller's idea that we will "change the earth's surface to suit us" was a Cold War–era interpretation of the biblical notion that man shall have dominion over nature and that the whole point of life on earth — and of the universe itself — is to advance the human race. In Teller's case, this translated into the desire to blast new harbors and redirect rivers, because, ultimately, the reason we have oceans or rivers is to increase human welfare — so why not reshape them? Of course, we now know that even if increasing human welfare was your only criterion, arbitrarily blasting harbors and redirecting rivers would not necessarily be a great way to advance the cause. But with Teller, the impulse was much bigger than that. The power of the bomb had made him godlike. He was not improving the world; he was remaking it to suit himself.

The dream of nuclear earthmoving—at least as Teller imagined it—is dead and gone. But as the world heats up, talk of moving rivers and mountains has returned. One physicist I spoke with predicted that northern rivers that flow into the oceans "with no purpose" would eventually be redirected to provide drinking water for the drought-parched south. Another scientist talked of blocking the Bering Strait to alter ocean currents. "If the sea levels are going to rise, why don't we create some massive inland lakes in Africa and Asia?" Sir Richard Branson, the wealthy British entrepreneur and founder of Virgin Atlantic Airways, wondered in a 2008 interview. "Instead of having all the cities flooded all over the world, the inland lakes can help take the brunt of it. At the same time, you would have this cool water, which would help cool the earth down. The water itself would help fertilize deserts, which would then grow trees."

Perhaps Teller's real brilliance was not as a scientist but as a provocateur. In his own perverse, indirect way, he articulated the central question of our time: what is the earth for? Is it a launching pad for human beings, a playground for us to explore before moving on to some greater destiny? Or is it a sacred and fragile place, a miracle of life in a cold and dead universe? The deeper you go into the logic of geoengineering, the more pressing this question becomes. Because as any engineer will tell you, before you can intelligently design something, whether it is a house, a bridge, or a climate system of a planet, the first question you have to ask is, what is its purpose?

....

FIVE

The Blue Marble

ON A SUMMER AFTERNOON in 2007, British scientist James Lovelock and I spent several hours in his slate-roofed cottage in Devon, England, talking about global warming and the dark future of human civilization. Then we stepped out the back door for a walk around Lovelock's thirty-five-acre property, where, as if to mock our fears, Mother Nature was showing some serious leg: honeybees buzzing, butterflies floating, a hawk circling overhead.

Lovelock's cottage is small and simple, surrounded by tall trees and hedgerows and approachable only by a labyrinth of roads that twist through the rolling green hills. At eighty-eight, Lovelock was still a jaunty walker. He and his wife, Sandy, usually put in five miles a day, rain or shine. During our walk, Lovelock kept up a lively commentary, pointing out the vines wrapping around old oak trees and a delicate but deadly flower called hemlock water dropwort. ("It's a nasty plant," he joked. "Al Qaeda would love it.") He's a small man, soft-spoken and polite, with white hair and large owlish glasses. Only his hands give him away as a scientist—a lifetime of handling chemicals has burned away the whorls on his fingertips.

Lovelock never whines about getting old, but he admits there are downsides. He's had forty operations, including a heart bypass and the removal of a kidney. He takes a handful of pills every morning and naps every afternoon. Because of a stomach condition, he has to be careful about what he eats. And, of course, his personal pleasures aren't what they were sixty years ago in London, when he was shagging nurses in underground first-aid stations while German bombs fell overhead. (When it comes to sex, Lovelock is refreshingly candid.)

Still, Lovelock has aged well. His eyesight is good. His hands don't shake. To get in and out of bed, he climbs an eight-foot ladder from his living room to his upstairs bedroom. His remarkable memory hasn't failed him — he can recall details from science papers he read sixty years ago. He travels constantly — in the months before my visit, he had attended conferences in Norway, Italy, the United States, and Australia. In England, he zooms around in his white Honda like a Formula One driver. He's up on the latest technology, from bots to BlackBerrys. Most important, Lovelock's theory of Gaia — first proposed in the late 1960s, which argues that the earth is best understood as a superorganism that is, in some sense, "alive" — is no longer dismissed as New Age quackery. "When Lovelock first wrote about Gaia forty years ago, his ideas were heresy," Ken Caldeira told me. "Now they are taught to schoolchildren."

But recently, Lovelock evolved from planetary scientist to doomsday prophet. After ignoring the issue of global warming for years — indeed, he was infamous for scorning those who talked about the "fragility" of nature — Lovelock is now convinced that the planet is in deep trouble. "Our future," he has written, "is like that of the passengers on a small pleasure boat sailing quietly above the Niagara Falls, not knowing that the engines are about to fail." In Lovelock's view, it doesn't matter how many rooftop solar panels we install or how tight we make the cap on greenhouse gas emissions — it's too late to stop the climate changes that are already under way. And those changes will be far more dramatic than people now suspect. By the end of the century, Lovelock believes, temperate zones such as North America and Europe could heat up by 17 degrees Fahrenheit, nearly double the high-end predictions of most climate scientists. Lovelock believes that this sudden heat and drought will set loose the Four Horsemen of the Apocalypse: war, famine, pestilence, and death. By 2100, he told me, the earth's population could be culled from today's seven billion to less than one billion, with most of the survivors living in the far latitudes—Canada, Iceland, Norway, and the Arctic basin.

I followed Lovelock through the trees and over a wooden bridge at the edge of his property. Below, the river was swollen by recent rains. The air was washed clean. I watched cows grazing on a distant hillside.

"On days like this," I said, "it's hard to worry about the future."

"It is," Lovelock agreed. Then he added, "But Gaia is pitiless, you know." His tone was cool, understated—not an angry preacher serving up a lecture, but a forthright doctor delivering an unwelcome diagnosis. "We've screwed her. And she will have her revenge."

Although Lovelock was born only a decade after Edward Teller, they were from two different worlds. To Teller, human beings were masters of the universe. To Lovelock, they are not much different from any other organism that has reached a population of seven billion. This is not to say that Lovelock equates humans with, say, fleas—he recognizes that an ability (and desire) to write books and build computers suggests we are indeed a higher order. But Mother Nature gives us no special dispensation just because we've produced Shakespeare, Einstein, and Johnny Cash. Like any other organism, if we grow beyond the carrying capacity of our natural habitats, if we despoil our nests, we will pay for it. As enlightened humans, our moral responsibilities are not just to each other but also to the health of the larger world we live in.

If you think this sounds like religion masquerading as science, you're not the only one. Type "Gaia" and "religion" into Google, and you'll get about three million hits—including Wiccans, spiritual travelers, massage therapists, and sexual healers, all deeply inspired by Lovelock's vision of the planet. One ministry declares that Gaia's "cunning mixture of science, paganism, eastern mysticism, and feminism have made this pagan cult a growing threat to the Christian Church." Ask Lovelock about those pagan cults, and he will grimace and tell you that he had no idea what a mess he was stepping into when he named his theory after the Greek goddess of the earth. Lovelock has no interest in softheaded spirituality or organized religion, especially when it puts human existence above all else. He once listened to Mother Teresa tell a crowd at the University of Oxford "to take care of the poor, the sick, and the hungry and leave God to take care of the earth." Outraged, Lovelock stood up and said, "I must disagree with the reverend lady. If we as people do not respect and take care of the earth, we can be sure that the earth, in the role of Gaia, will take care of us and, if necessary, eliminate us."

Lovelock's fears about the future of human civilization are all the more alarming because he is not an alarmist by nature. In his view, the dangers of nuclear power are grossly overstated. Ditto the dangers of mercury emissions in the atmosphere, the genetic engineering of food, and the loss of biodiversity on the planet.

When it comes to global warming, what distinguishes Lovelock from other scientists is how he defines the problem. Lovelock has no quibble with the relationship between rising greenhouse gas pollution and the warming of the planet. What he balks at is the idea that if we cut back CO_2 emissions by, say, 80 percent over the next few decades—a difficult if not impossible goal on its own, Lovelock believes—that will solve the problem. "It won't make a damn bit of difference," he told me. "If there were only a billion people living on the planet, we could do whatever we please. But there are nearly seven billion. At this scale, life as we know it today is not sustainable."

In Lovelock's view, Gaia's autopilot system—the giant, inexpressibly subtle network of positive and negative feedbacks that keep the earth's climate in balance—is out of whack. Pollution is just one cause. Changes in land use are just as important, if not more so. Plants, which suck up CO_2 and exhale oxygen through photosynthesis, are a key part of the global carbon cycle. Knocking down rain forests and converting the land to agriculture is the equivalent of cutting out the lungs of the planet. It's not the fate of the earth that's at stake; it will eventually recover its equilibrium, even if it takes millions of years. What may not survive is human civilization. As Lovelock put it, "You could quite seriously look at climate change as a response of the system designed—well, 'designed' would be a very dangerous word to use—*intended* to get rid of an irritating species: us humans. Or at least cut them back to size."

Lovelock believes that most scientists, politicians, and environmentalists have failed to grasp—much less prepare us for—the urgency and consequences of global warming. Some of this failure, Lovelock argues, is just plain ignorance. Some of it is due to misplaced faith in computer models as reliable guides to future climate change. Some of it is due to a Christian faith that God will take care of us. Some of it is greed. (Lovelock has reserved a special place in hell for cleantech entrepreneurs who want to get rich by "saving" the planet.) And some of it is simply cowardice, or a fear of being proved wrong.

Lovelock himself has no such fears. In his view, it is time to put away sentimental notions about nature and focus our attention on doing whatever it takes to make sure civilization survives Gaia's coming rage. And if that means trying to modulate that rage by brightening clouds or throwing some particles into the stratosphere to block out sunlight, so be it. Although Lovelock views the idea of large-scale geoengineering as an act of profound hubris—"I'd sooner expect a goat to succeed as a gardener than expect humans to become stewards of the earth"—he also thinks it may be necessary as an emergency measure, much like kidney dialysis is necessary to a person whose health is failing. "At best," he told me, "perhaps it will buy us some time."

Lovelock is a child of the industrial age. He grew up with soot in his lungs, coughing and pale and poor in South London. He was an only child, conceived, he believes, in a moment of joy over the surrender of the German army on Armistice Day in 1918. ("The math is right," Lovelock joked.) His mother was an early feminist who devolved into a frustrated, embittered housewife. His father grew up so desperately hungry that he spent six months in the notorious Reading prison when he was fourteen for poaching a rabbit from a local squire's estate. He later worked at a gas plant and, for a time, ran a small art gallery in a working-class London neighborhood.

Shortly after Lovelock was born, his parents passed him off to his grandmother to raise. "They were too poor and too busy to raise a child," Lovelock said somewhat cryptically. "For all emotional and practical purposes, my grandmother was the mother figure of my childhood." In school, he was a lousy student — mildly dyslexic and more interested in pranks than homework. But he loved books, especially the science fiction of Jules Verne and H. G. Wells.

To escape the soot of urban life, Lovelock's father sometimes bundled him onto a train or bus and traveled out into the countryside, where they took long walks together. (He remained close with his parents even though they gave him away.) His father, who was a hunter, knew all the animals' tracks and all the names of the wildflowers. Lovelock caught trout by hand from the streams and gorged on blueberries. Later, he spent a lot of time biking through the countryside on his own, exploring rivers and valleys. The freedom and romance he felt on these jaunts outside the city had a transformative effect on him. "It's where I first saw the face of Gaia," he told me.

By the time Lovelock hit puberty, he knew he wanted to be a scientist. His first love was physics. But his dyslexia made complex calculations difficult. Instead, he opted for chemistry. After high school, he took a day job in a lab and attended Birkbeck College (part of the University of London) at night. The year after he enrolled, Germany invaded Poland. Instead of joining the military, Lovelock converted to Quakerism and became a conscientious objector. In his written statement, Lovelock explained why he refused to fight: "War is evil."

Out in the real world, Lovelock quickly demonstrated a talent for scientific problem solving. He took a job at the National Institute for Medical Research in London, where one of his first assignments was to develop new ways to stop the spread of infectious diseases, which many health officials were concerned about in war-torn England. Lovelock spent months in underground bomb shelters studying how viruses are transmitted and ended up inventing the first aerosol disinfectant. A few years later, he became interested in a new scientific frontier — cryogenics, or cold-temperature biology. He was the first to understand how cellular structures respond to cold temperatures and developed a means to freeze and thaw animals and their sperm that is still in use today.

But Lovelock's most important invention was the electron capture detector (ECD). In 1957, while working with a gas chromatograph, an instrument used to measure the chemicals in gases and other materials, Lovelock hacked together a device to measure minute concentrations of certain reactive gases in the air, such as pesticides and chlorofluorocarbons (CFCs). The instrument, which he literally built at his kitchen table, fit into the palm of his hand and was so exquisitely sensitive that if you dumped a bottle of some rare chemical on a blanket in Japan and let it evaporate, the ECD would be able to detect it in England a few weeks later. The device was eventually redesigned by Hewlett-Packard, which ended up selling tens of thousands of them for industrial use. If Lovelock had bothered to get a patent for the device, he would have been a rich man. But he has never cared much for money, except to buy himself the freedom to remain an independent scientist.

As it turned out, Lovelock's invention of the ECD roughly coincided with the publication of *Silent Spring* in 1962, which alerted the world to the dangers of pesticides such as DDT and other chemicals. Although Rachel Carson made no mention of the ECD in her book (Lovelock says they never met or talked), by the time her book appeared scientists were already using the device to measure the residue of pesticides in the fat of Antarctic penguins and in the milk of nursing mothers in Finland, giving hard evidence to Carson's claims that man-made chemicals were indeed impacting the environment on a global scale.

A decade or so later, Lovelock made a discovery that was even more important than his invention of the ECD. While staying at a small vacation house he'd purchased in far western Ireland in the late 1960s, he took a random sample of the haze that drifted into this remote area and found it laced with CFCs (see chapter 2), a sure sign of man-made pollution. Lovelock thought, If CFCs are in remote Ireland, where else might they be?

Lovelock hitched a ride on a research vessel for a six-month voyage to Antarctica, where he used a jury-rigged ECD to detect the buildup of CFCs, as well as other man-made chemicals, in the atmosphere. Lovelock wrote up his discoveries in three separate papers, demonstrating that man-made chemicals were now ubiquitous on the planet. It was an important insight, but Lovelock failed to grasp the dangers these CFCs posed to the earth's atmosphere, stating in a scientific paper that this buildup posed "no conceivable hazard."

It was, as Lovelock said to me, "one of my greatest blunders." It was a mistake that may have cost him a share in a Nobel Prize.

As it turned out, CFCs weren't toxic to breathe, but that didn't mean they didn't pose an even larger hazard to people. In 1974, scientists F. Sherwood (Sherry) Rowland and Mario Molina hypothesized that CFC molecules could be split apart by sunlight, creating chlorine atoms that would burn a hole in the protective ozone layer of the atmosphere. This would in turn allow dangerous levels of ultraviolet light to reach the earth. Sure enough, such a hole was soon found over Antarctica, and the ozone hole became a symbol of humankind's potential to unknowingly trash the planet on a grand scale. CFCs were banned, ozone depletion was halted, and in 1995 Rowland and Molina shared a Nobel Prize.

"If Lovelock hadn't detected those CFCs," said Stanford University biologist Paul Ehrlich, "we'd all be living under the ocean in snorkels and fins to escape that poisonous sun."

In 1961, at about the same time Edward Teller was dreaming of using nuclear bombs to dig harbors and move mountains, Lovelock was working at the National Institute for Medical Research's Mill Hill laboratory in North West London. It was a good job, offering decent pay and plenty of freedom, but he was bored. The ECD was a nice invention, but neither Lovelock nor the rest of the world yet appreciated the significance of it. He was forty-two years old, the father of four children (including the youngest, John, who was born with a birth defect that left him brain-damaged). Lovelock seemed to be on the fast track to scientific mediocrity.

One day a letter from NASA's director of Space Flight Operations arrived in Lovelock's mailbox, inviting him to join a group of scientists who were about to explore the moon. Lovelock had never heard of NASA, but some of his inventions, including the ECD, had come to the attention of NASA administrators. They believed that his skills might be used to help them design and build research instruments for the Apollo space program. Within a few weeks, Lovelock had quit his job, packed up his family, and moved to California to join the space race.

He didn't have to spend much time at NASA's Jet Propulsion Laboratory in Pasadena, however, before realizing that, scientifically speaking, the moon wasn't a very interesting place. The real excitement was Mars. "With the moon, the question was, Is it safe for astronauts to walk on the surface?" Lovelock explained to me. "With Mars, the question was, Is there life there?"

Lovelock's colleagues at JPL struggled to design instruments to test for life, or lifelike substances, on the planet's surface. One eminent biologist showed Lovelock a cage of stainless steel, about one centimeter long, that he had built and proposed sending on a mission to Mars. When Lovelock asked what it was, the biologist replied, "A flea trap." The biologist explained that Mars was a great desert, and in any desert, there are sure to be fleas.

Lovelock had a better idea: why not analyze the chemical composition of the Martian atmosphere? If there were life on Mars, Lovelock reasoned, those organisms would be obliged to use up the raw materials of the atmosphere (such as oxygen), while at the same time dumping waste products (such as CO_2), just as life on earth does. Even if the materials consumed and discharged were different, the imbalance in the chemical equilibrium of the Martian atmosphere would be relatively simple to detect.

But there was no imbalance in the Martian atmosphere. Using an infrared telescope, scientists determined that the atmosphere on Mars was close to chemical equilibrium and dominated by CO_2. (With the right instruments, the different wavelengths of light absorbed by various chemicals can be seen in the spectrum of light reflected from the planet.) In comparison, the earth's atmosphere is in a state of deep disequilibrium, where CO_2 is only a trace gas and there is an abundance of reactive gases, including oxygen and methane. Why? Because earth has life on it. Mars, Lovelock concluded, was probably dead. But if life was creating the earth's atmosphere, Lovelock reasoned, it must also, in some sense, be regulating it. Lovelock knew, for example, that the sun has been growing steadily hotter over the eons. (It is now about 25 percent hotter than when life began on the planet.) What was modulating the surface temperature of the earth, keeping it cool enough for life? To put it another way, whose hand was on the thermostat?

The concept of the earth as a kind of superorganism was not new. In 1490, Leonardo da Vinci believed pretty much the same thing. In the 1930s, British ecologist Arthur Tansley coined the term "ecosystem" to suggest the interconnecting networks of life. A decade or so later, Norbert Wiener introduced the term "cybernetics," an influential idea that described the control and communication of animals and machines through a network of feedback systems. And in 1972, the famous "blue marble" photograph taken from *Apollo 17* as it circled twenty-seven thousand miles above the earth inspired many people to think of our planet as a single entity—fragile, vulnerable, isolated. But it was Lovelock who put all these ideas together and realized that thinking of the earth as a superorganism was an extremely useful and provocative concept. Lovelock soon abandoned his work with NASA and moved back to England, where he discussed his new idea with his neighbor, novelist William Golding (*Lord of the Flies*), who suggested that Lovelock needed a catchy name. Why not Gaia?

Lovelock began a collaboration with American biologist Lynn Margolis, who helped him refine his idea of the earth as a self-regulating system, especially the role microorganisms play in creating atmospheric gases such as methane. Lovelock tried to publish his ideas about Gaia in mainline scientific journals, but nobody would touch the topic. So he ditched the scientific establishment altogether and set his ideas out in his first book, *Gaia: A New Look at Life on Earth,* published in 1979. "The Gaia hypothesis," Lovelock wrote, "is for those who like to walk or simply stand and stare, to wonder about the earth and the life it bears and to speculate about the consequences of our own presence here. It is an alternative to that ... depressing picture of our planet as a demented spaceship, forever traveling driverless and purposeless around the inner circle of the sun."

Hippies loved it; Darwinists didn't. Richard Dawkins, author of *The Selfish Gene,* dismissed Lovelock's book as "pop ecology literature." British biologist John Maynard Smith went further, calling Gaia "an evil religion." In their view, Lovelock's concept of the earth as a superorganism flew in the face of evolutionary logic: If the earth is an organism, and organisms evolve by natural selection, then that implies that somehow the earth has outcompeted other planets. How is that possible? Also troubling was Lovelock's suggestion that the earth is kept comfortable by and for its inhabitants—you and me and the creatures that live with us. It was obvious to other scientists that geophysical processes—oceans, clouds, the weathering of rocks—also are involved. More to the point, by arguing that life creates the condition for life, Lovelock seemed to be suggesting that there was some predetermined purpose or goal. In the minds of many of his peers, Lovelock was dancing very close to God—or at least to intelligent design.

But that was not what Lovelock had in mind at all. Large systems, in his view, don't need a purpose. To prove it, he and a colleague, Andrew Watson, devised a simple, elegant computer model called Daisyworld. Their intent was to show how organisms evolving under rules of natural selection are part of a system that is self-regulating.

Imagine an earthlike planet where only black and white daisies grow and where the sun gets slowly hotter over time. By their nature, black daisies absorb the sun's heat, whereas white daisies reflect it back into space. In the early years of the planet, when the atmosphere was very cool and the sunlight not strong, small colonies of black daisies would flourish at the equator. By absorbing heat, the black daisies would form warmer spots where still more black daisies would grow. Before long, white daisies would also start to take advantage of the favorable temperature. Slowly, the sun's heat would increase to the point where the black daisies at the equator would get too hot and start to die off. The black daisies would then retreat to the poles. As the sun continued to heat up, the white daisies at the equator would flourish, reflecting the sun's heat out of the atmosphere and helping to maintain a temperature suitable for daisy life. In this experiment, the daisies had no purpose. They were not acting for the good of all; they were not altruistic or conscious. They simply existed, and by existing, they altered their environment.

Daisyworld quieted some of Lovelock's critics, but the scientific debate about his theory raged for most of the 1980s. Lovelock continued refining his thoughts despite troubles in his personal life. His wife, Helen, who had been diagnosed with multiple sclerosis in her forties, was in the midst of a slow and painful decline. She died in 1989. (He married his current wife, Sandy, shortly afterward.) Lovelock himself had several major surgeries, including the removal of a kidney after it was damaged in a tractor accident in Devon. He supported himself largely with consulting gigs for Hewlett-Packard, which was interested in his expertise in scientific instruments, and for MI5, the British security service, where he often put his chemistry background to use helping the agency analyze explosives used by the Irish Republican Army.

Among scientists, Lovelock greatly redeemed himself with a second book, *The Ages of Gaia,* which offered a more rigorous exploration of the feedback mechanisms that keep the earth's atmosphere suitable for life. Plankton in the oceans, for instance, help cool the planet during warm times by giving off dimethyl sulfide, a chemical that oxidizes into tiny particles that cause the formation of clouds above the oceans. The clouds reflect the sun's heat back into space. "In the 1970s, plenty of us thought Gaia was nonsense," said Wally Broecker, a pioneering paleoclimatologist at Columbia University. "But Lovelock got everyone thinking seriously about feedback and the dynamic nature of the planet." Of course, scientists like Broecker rarely use the word "Gaia." They prefer the phrase "Earth System Science," which views the world, as one treatise defined it in 2001, as "a single, self-regulating system comprised of physical, chemical, biological and human components." In other words, Gaia in a lab coat.

But the ultimate power of Lovelock's vision transcends science. "What makes . . . the Gaia Hypothesis so inspiring?" former Czech president and Nobel Prize winner Václav Havel asked in a 1994 talk.

One simple thing: [it] reminds us, in modern language, of what we have long suspected, of what we have long projected into our forgotten myths and perhaps what has always lain dormant within us as archetypes. That is, the awareness of our being anchored in the earth and the universe, the awareness that we are not here alone nor for ourselves alone, but that we are an integral part of higher, mysterious entities . . . This forgotten awareness is encoded in all religions. All cultures anticipate it in various forms. It is one of the things that form the basis of man's understanding of himself, of his place in the world, and ultimately of the world as such.

Lovelock understands as well as anyone that geoengineering is a dangerous idea. Consider the troubles he has had managing his little spread in Devon, much less the entire planet. When he bought the place thirty years ago, it was surrounded by fields shorn by a thousand years of sheep grazing. But to Lovelock, open land reeks of human interference with Gaia, so he set out to restore his thirty-five acres to its more natural character. After consulting with a forester, he planted (with help) twenty thousand trees — alders, oaks, pines. Unfortunately, he planted many of them too close together, and in rows. The trees are about forty feet tall now, and rather than feeling "natural," parts of his land have the look of a badly managed forestry project. "I botched it," Lovelock said with a grin as we hiked through the woods together. "But in the long run, Gaia will take care of it."

None of that "fragile web of life" nonsense for Lovelock. Trees grow back. Species vanish, and new ones are born. Cities rise and fall. Even global warming, he believed until recently, would be little more than a passing fever. So what changed his mind?

In 2004, Lovelock's friend Richard Betts, a researcher at the Met Office Hadley Centre, Britain's top climate institute, invited him to stop by for a day of tours and talks with the scientists there. Lovelock went from meeting to meeting, hearing the latest data and modeling results about the melting of polar ice, the decline of tropical rain forests and boreal forests, and the carbon cycle in the oceans. "It was terrifying," Lovelock told me. "We were shown five separate scenes of positive feedback in regional climates — polar, glacial, boreal forest, tropical forest, and oceans — but no one seemed to be working on whole-planet consequences." Equally chilling, Lovelock said, was the tone in which the scientists talked about the changes they were witnessing, "as if they were discussing some distant planet or a model universe, instead of the place where we all live."

As he was driving home that evening, it hit him. The resiliency of the system was gone. The forgiveness had been used up. "The whole system," Lovelock recalled thinking, "is in failure mode."

To Lovelock, the fact that nobody at the Hadley Centre quite realized how much trouble we are in was not surprising. He had long been critical of mainstream scientists for thinking too much about their narrow areas of expertise and missing the big picture. "Scientists are now like medieval clerics," he said. "They're all concerned about the number of enzymes that can dance on the head of a pin." Equally problematic in Lovelock's view is the reliance on computer climate models to predict our future. If the models don't predict catastrophe, why worry? Because models are not simulators, for one thing. Even the best models leave a lot out, such as cloud physics and the dynamic response of ecosystems to climate changes. And Lovelock knows better than most people how dangerous it is to depend on computer models for predictions about something as complex as the earth's climate. In the 1980s, it was modelers who assured everyone that ozone depletion was no cause for concern. Then some intrepid scientists went down to the Antarctic and took some actual measurements and discovered a huge—and potentially deadly—ozone hole. It turned out some equations in the models were wrong.

Could the models used to calculate the impact of rising greenhouse gas emissions also be wrong? Take the projected rise in sea level that I mentioned in chapter 1. In a 2007 report, the IPCC estimated that by 2100, global warming due to rising greenhouse gas emissions would cause inland glaciers to melt and seas to expand, causing a maximum sea level rise of only twenty-three inches. That report assumed little or no contribution to twenty-first-century sea level rise from Greenland or Antarctica. Greenland, the report said, would take a thousand years to melt.

But new data shows that the polar ice sheets are melting far more quickly than computer models suggest, and some of the very same scientists who contributed to the 2007 IPCC report now believe that the sea could rise by three feet or more by the end of the century. As I noted in chapter 1, NASA's James Hansen takes it even further, arguing that the seas could rise by as much as nine feet.

Why do computer models say one thing and the real world another? "Because modelers don't have the foggiest idea about the dynamics of melting ice sheets," Lovelock explained.

It's not just ice that the models have trouble with. Clouds are also problematic. In particular, satellite data has shown that cloud convection data in the tropics is misrepresented in most climate models. And although it is certainly true that the models could be flawed in either direction, Lovelock believes that the observational data in recent years makes it clear that climate models have been too conservative. "There are plenty of scientists who know exactly how flawed these models are and how dire our situation has become," Lovelock said. "A number of them are willing to talk with me about it in private, but they are unwilling to discuss it in public. They are afraid that if they are accused of being alarmists, they'll lose their jobs."

The upshot of all this, in Lovelock's view, "could be cataclysmic." To grossly oversimplify the doomsday scenario: Rising heat means more ice melting in the poles, which means more open water and land, which in turn increases the heat (ice reflects sunlight; open land and water absorb it), causing more ice to melt and sea levels to rise. More heat pumps up the hydrologic cycle, leading to more intense rainfall in some places, droughts in other. The great boreal forests die. The Amazon rain forests wither. The ocean algae system weakens. The permafrost in northern latitudes thaws, releasing methane, a greenhouse gas that is far more potent than CO_2 ... On and on it goes.

In a functioning Gaian world, these positive feedbacks are modulated by negative feedbacks, the largest of which is simply the earth's ability to radiate heat into space. But at a certain point, Lovelock argues, the regulatory system breaks down, and the earth's climate makes a jump—as it has many times in the past—to a new, hotter stable state. Scientists have noted the same kind of instability in other complex systems. Overfishing, for example, tends to cause increased fluctuations in fish populations. At a certain point, if the fishing continues, the populations reach a tipping point, the fluctuations cease, and the fish disappear. Financial markets behave in a similar way before a crash, and so do brain waves before an epileptic seizure. If the earth's climate were to flip to some new hotter state, it would not be the end of the world. The earth has survived many such dramatic transformations. But it certainly would be the end of the world as we know it. "Nobody likes to think about how fragile civilization really is," Lovelock told me over lunch one day. "We really could lose it all."

Let's assume that Lovelock is right and we are indeed poised above Niagara Falls. What can we do? In Lovelock's view, modest cuts in greenhouse gas emissions won't help us. That doesn't mean it isn't worth cutting emissions—"after all, I may be wrong about all this," he admits—but he believes that the climate is already moving and we're not going to stop it by throwing solar panels up on the roof of every Wal-Mart. What about capturing CO_2 pollution from coal plants and pumping it underground? "We can't possibly bury enough to make any difference." Carbon trading? "Goldman Sachs is going to make a lot of money off it, but it's meaningless." Biofuels? "A monumentally stupid idea." Renewables? "Won't make a dent." To Lovelock, the whole idea of sustainable development is wrong-headed: "We should be thinking about sustainable retreat."

Lovelock isn't talking about cutting back on our shopping trips to Target. He's talking about changing how we live and where we get our food; about making plans for the migration of millions of people from low-lying regions such as Bangladesh into Europe and Asia; about admitting that New Orleans is a goner and moving people to cities better positioned for the future. Most of all, Lovelock says, it's about everybody "absolutely doing their utmost to sustain civilization, so that it doesn't degenerate into dark ages, with just warlords running things, which is a real danger."

"I wish I could say that wind turbines and solar panels will save us," Lovelock told me. "But I can't. I think people need to hear the truth about what's going on. And I'm certainly not suggesting we do nothing. Quite the opposite. We need bold action. We have a tremendous amount to do."

In Lovelock's view, we have two choices: we can live in equilibrium with the planet as hunter-gatherers, or we can live as a very sophisticated high-tech civilization. There is nothing in between. "There's no question which path I'd prefer," he said one morning in his cottage, grinning broadly and tapping the keyboard of his Dell computer. Despite his hermitlike existence, Lovelock is an unabashed technophile, a man who believes that if we are bold enough and visionary enough, we can engineer ourselves out of any nightmare. "It's really a question of how we organize society—where we get our food, water, how we generate energy."

For water, the answer is pretty straightforward: desalinization plants, which can turn ocean water into drinking water. In many regions, the food supply will be a more complex problem. Heat and drought will devastate many of today's food-growing regions. These problems also will push people north, where they will cluster in cities. In these areas, there will be no room for backyard gardens. Today's global agricultural network will be long gone. The most likely result, Lovelock believes, is that we will synthesize food—grow it in stainless steel vats from tissue cultures of meats and vegetables. It sounds far-out and deeply unappetizing (and all too suggestive of *Soylent Green*, the sci-fi classic about a polluted, overpopulated world in which the main food staple is a wafer that turns out to be made from human beings). But technologically speaking, growing food in vats wouldn't be hard to do.

A steady supply of electricity will also be vital. Five days after his visit to the Hadley Centre in 2004, Lovelock penned a fiery op-ed for the *Independent*, a major British newspaper, titled "Nuclear

Power Is the Only Green Solution." Lovelock argued that we should "use the small input from renewables sensibly," but that "we have no time to experiment with visionary energy sources; civilization is in imminent danger and has to use nuclear—the one safe, available, energy source—now or suffer the pain soon to be inflicted by our outraged planet."

Then there's geoengineering. When Lovelock talks about taking emergency measures to cool the planet, he often uses medical analogies, comparing the earth to a cancer patient whose only chance of survival is to undergo a dangerous chemotherapy treatment. (Lovelock isn't the only one who thinks this way. In a 2006 interview, Al Gore said, "The earth has a fever and just like when your child has a fever, maybe that's a warning of something seriously wrong.") The analogy of the sick patient, of course, brings up the question of who the doctor is. The trouble, in Lovelock's view, is not that the earth is such a complex system. "As a system, I'd say it's a good deal less complex than the human body," he told me. The problem is that we have so little experience trying to fix it. "Our knowledge of the earth's system is about where our knowledge of the human system was at the turn of the century," Lovelock said. "We have so much to learn—but so little time to learn it."

And in the view of many critics of geoengineering, it's precisely because we know so little about the system we're trying to manipulate that we shouldn't try. "It's not like chemotherapy—it's like Russian roulette," one climate scientist has argued. "We have no idea what is going to happen if we start messing around with this."

"I agree," Lovelock told me one afternoon. "We're very likely to screw it up. But if things get bad, we may not have any choice. At the very least, we should prepare. We should learn all we can about what we can and can't do. Ignorance is not our friend." In other words, when it comes to geoengineering, it might be smart to begin sorting good ideas from bad, lest we fall under the spell of another generation of Charles Hatfields. It was essentially the same argument I'd heard from David Keith and others, and one that I found increasingly difficult to dispute. "Before we count on this as a backup plan, we need to do some serious research on what might work and what might not," said David Victor, a political scientist at the University of California, San Diego, who has written several influential articles about the political implications of geoengineering. "The time to discover the flaws in these geoengineering ideas is now, not in the midst of a political emergency."

The tricky part of the argument, of course, is where you draw the line between learning and doing, especially given the flaws and faults of computer models. At some point, the only way to really learn about the strengths and weaknesses of various geoengineering schemes is to give them a try.

Despite his gloomy forecast, Lovelock clearly loves the challenge of coming up with ways to tinker with the planet. He has always been as much an inventor as a scientist, the kind of man who believes that mysteries are most often revealed through new gadgetry. When I visited him in 2007, he was intrigued by biochar, a material made by "cooking" organic matter such as farm waste in a low-oxygen stove so that carbon is not released into the atmosphere. The charcoal-like material that is left over is then buried, simultaneously enriching the soil and storing the carbon. Lovelock was also noodling around with ways to manipulate the oceans, including injecting zillions of tiny air bubbles to increase their reflectivity. The natural carbon-sucking ability of the oceans fascinates him, too. One thought: suspend two-hundred-foot vertical pipes, each outfitted with a valve at the bottom, into the tropical oceans and allow deep, nutrient-rich water to be pumped to the surface by wave action. Nutrients from the deep water would increase algae bloom, which would absorb CO_2 and help cool the planet. Critics have pointed out that even if the pipes worked, it would take tens of thousands of them to make even a small difference. But that hasn't deterred Lovelock. "It's a way of leveraging the earth's natural energy system against itself," he mused. "I think Gaia would approve."

Lovelock's ocean pipes are unlikely to be any more effective against global warming than leeches would be against leukemia. But the premise of Lovelock's vision is indisputable: if the dominant metaphor of our time is of the earth as a sick child (or a cancer patient), then it is inevitable that we will try to cure that child — especially since our survival depends on it. To some, the idea of the earth as a self-regulating system not unlike the human body is both overly simplistic and frighteningly mechanistic. But there is another way to look at it, one that harks back to early Enlightenment thinkers who believed that what we call "nature" is in fact a collection of rules and forces that can be studied, understood, and, possibly, manipulated. In many ways, Lovelock's vision is an antidote to the chaos of twentieth-century science, which fragmented the world into quarks, quantum mechanics, and untouchable mystery.

In any case, there is no turning back now. "Like it or not, we are the brains and nervous system of Gaia," Lovelock told me as we hiked through the underbrush on his property. "We have now assumed responsibility for the welfare of the planet. How shall we manage it?"

SIX

Doping the Stratosphere

FOR AMERICAN SCIENTISTS, life in the gulag ended November 4, 2008, with the election of President Barack Obama. After eight years of faith-based reasoning and political voodoo, America had a president who trusted rational thought and scientific inquiry. The impact was immediate and profound. A few weeks after the election, at the annual meeting of the American Geophysical Union (AGU), where earth and space scientists gather to discuss the latest research, you could actually feel the beginnings of a scientific renaissance. During the conference, the bright, open atrium of the Moscone Center in San Francisco was buzzing with energy—the fifteen thousand scientists in attendance were twittering, blogging, emailing, or chatting among themselves about the impact of drought in California, the persistence of CFCs in the atmosphere, or ice sheet hydrology. "Science matters again," one attendee told me, as if it were a great surprise.

But from a global warming point of view, the news from the meeting was anything but inspiring. James Hansen warned a packed auditorium about the dangers of runaway warming, suggesting that if we burn our entire reserves of coal and oil and tar sands, we could boil the oceans away and turn the earth into Venus—a hot rock swathed in CO_2 and devoid of life. Eric Steig of the University of Washington in Seattle presented important research showing that, contrary to the talking points of some climate skeptics, Antarctica has warmed significantly over the past fifty years. Other talks covered the consequences of a diminished snowpack in the Sierra Nevada on California drinking water supplies, the effects of Australian wildfires on regional biodiversity, and a risk analysis of asteroid impacts versus climate change (guess which one is the greater threat).

It was precisely this kind of research—the ever-accelerating accumulation of evidence that global warming is happening now and happening fast—that convinced a group of concerned scientists to hold three private dinners during the AGU meeting to consider how to advance our knowledge of the risks and benefits of geoengineering and, more specifically, what might be done in the case of a climatic emergency. In a sense, the dinners were a manifestation of Lovelock's view that it is time to start thinking hard about a Plan B to save civilization. What is the best way to cool the planet in a hurry? What kind of research needs to be done in order to be prepared? Who should lead the research? When should the first tests begin?

The dinners were held in the private dining room of a fashionable restaurant called Shanghai 1930, which is only a few blocks away from the Moscone Center. They were hosted by David Keith and Ken Caldeira, with about twenty people in attendance each night. All of the attendees were scientists, except for me, a Silicon Valley entrepreneur, and a fellow from the American Enterprise Institute, a conservative think tank in Washington, D.C. The one thing everyone had in common was a belief that questions about geoengineering are going to be an important part of the debate over how to deal with global warming in the next decade or so, and it was time to begin separating fact from fiction.

But in a larger sense, the dinners represented a turning point in the evolution of geoengineering as a policy tool—the moment when the conversation moved from "Can we do it?" and "Should we do it?" to the much more focused "*How* do we do it?" Or, as Caldeira put it during his introductory remarks shortly after the first dinner began, "If you are pushed against the wall in a Senate meeting room and asked what you can do to cool off the planet in a hurry, what do you say?"

Caldeira stood at the head of the table in jeans and a dark blue pullover, presiding over the group like an absent-minded professor in front of a bunch of restless students. He is in his early fifties, with mischievous eyes and unkempt dark, curly hair. He spoke to the scientists in a halting, distracted way, his mind obviously wandering from idea to idea. He often paused to interject a witty aside ("This meeting may be nefarious, but it's not secret") or to grapple with unexpected insights that popped into his head. I had met him at the very beginning of my research for this book and quickly discovered that he is one of the few people who has a dynamic model of the entire earth's climate in his head and can talk about it in plain English. As a scientist, he has done groundbreaking work on ocean acidification and the long-term impacts of high levels of CO_2 in the atmosphere. But he also was a prominent antinuke activist back in the 1980s, plays bass in a rock band, is married to a tall, glamorous Russian woman, and commutes to his office on a Vespa scooter. (On rainy days, he drives a black BMW that, until recently, had a sticker on the rear bumper that said JAIL BUSH.) And in a world fogged with complex data and abstruse theories, Caldeira has a way of cutting through the crap. For example, he likes to compare emitting CO_2 to mugging old ladies: "It's wrong to mug little old ladies, and it's wrong to emit carbon dioxide into the atmosphere," he told me. "The right target for mugging little old ladies and for carbon dioxide emissions is zero."

The better I got to know Caldeira, the more I trusted and respected him. Still, as I sat at the table that first night, I couldn't escape the feeling that I was trapped on the set of a sci-fi movie. Here we were, in a dark, quiet room with Asian art and velvety red walls, sampling hors d'oeuvres from circulating waiters, debating how many bottles of wine to uncork, and discussing what could be done to stop the melting of the planet. Yes, Caldeira and all the other scientists in the room are thoughtful, ethical, well-intentioned people who were well aware that they were in uncharted waters. But that self-awareness didn't change the fundamental fact that what they were discussing was, on some level, insane.

For me, this feeling was underscored by the presence of a big, barrel-chested, red-bearded, sixty-five-year-old physicist in a tie-dyed sweatshirt. I had met Lowell Wood several years earlier, when I interviewed him for a story I wrote for *Rolling Stone.* Wood is a dark star in the world of science, a larger-than-life character who was a protégé of Edward Teller, with whom he worked at the Lawrence Livermore National Laboratory for nearly four decades. During the 1980s and 1990s, as Teller aged and withdrew, Wood became one of the Pentagon's top weaponeers, its go-to guru for threat assessment and weapons development. Wood is infamous for championing fringe science, from x-ray lasers to cold fusion nuclear reactors, as well as for his long affiliation with the Hoover Institution, a right-wing think tank on the Stanford campus. To some scientists, Wood is a brilliant outside-the-box thinker. To others, he is the embodiment of Big Science gone awry.

Wood's presence at the dinner was portentous in several ways, not least because it was a sign that geoengineering had not entirely emerged from the shadows of the Cold War. But if anyone deserved a spot at this dinner table, it was Wood. For one thing, as Caldeira had told me earlier, "no one has done more thinking about the nuts and bolts of geoengineering than Lowell." Since retiring from Livermore in 2007, Wood had moved to Seattle, where he had gone to work with his old friend Nathan Myhrvold, the former chief technology officer at Microsoft, at a company called Intellectual Ventures. Myhrvold's goal at his new company, Malcolm Gladwell wrote in *The New Yorker,* "was to see whether the kind of insight that leads to invention could be engineered." On a more practical level, the basic business plan at Intellectual Ventures is to buy patents and license them to interested companies. In Seattle, Wood remained deeply involved in geoengineering, exploring techniques for redirecting hurricanes, among other things, but also serving as an intellectual broker between academic scientists and entrepreneurs in the Myhrvold/Microsoft orbit.

One of the people in that orbit was Bill Gates. Since announcing a few years earlier that he was leaving his position as chairman of Microsoft, Gates had turned his attention to the Bill & Melinda Gates Foundation, a private philanthropy devoted to, among other things, reducing poverty and disease in the developing world. But, as Gates knows as well as anyone, you can't talk about the future of the developing world without talking about drought, rising seas, and changing disease vectors, and although Gates is no tree-hugger, he found himself increasingly forced to confront the problem of global warming. Through Myhrvold, Gates met Wood, with whom he informally discussed a variety of issues, including the science of global warming and the future of nuclear power. (In my conversation with Gates, he jokingly referred to Myhrvold and Wood as "reasonably crazy people.") In late 2006, Gates also began a series of informal meetings with Keith and Caldeira to discuss a variety of topics related to climate science and energy policy, including geoengineering.

Like any rational, well-informed person, Gates understood very well that geoengineering is not an idea that we should rush to embrace. But he was also intrigued by the notion, in part because he respected and trusted Caldeira and Keith, and in part because he is interested in technological solutions to difficult human problems. Although Gates clearly believes that we can find a way to power modern life without cooking the planet, he is impatient with the slow pace and gross underfunding of clean energy research and development, especially given recent news of melting ice and increasing droughts. ("Everyone is afraid as hell," he told me. "And they should be afraid as hell.") In Gates's view, the obvious question is, is there enough time to remake the world's energy system and still avert dangerous climate change? "In the world of energy," he said, "innovation takes fifteen years, deployment takes twenty years. So if something like geoengineering can buy you twenty years or thirty years, then maybe that's a good thing."

Keith and Caldeira agreed that research into the risks and benefits of geoengineering was hamstrung by the fact that there was very little money available to explore even the most basic questions. Gates offered to help with that, eventually giving them a few million dollars in philanthropic funding (which means, essentially, that he had no control over how they spent it, nor could he profit from whatever technologies they might develop) to advance their research on geoengineering and other climate- and energy-related topics. It was Gates's money, in fact, that was paying for the dinners at the AGU meeting in San Francisco.

Discussion during the dinners was straightforward and provocative. There was much talk about climate emergencies, droughts and famines, and what might trigger a call for quick deployment of geoengineering schemes. It was generally agreed that the most likely near-term catastrophe was the continued acceleration of warming in the Arctic, which could not only result in rising sea levels and major changes in weather patterns in the Northern Hemisphere but also could cause a massive release of methane gas from the thawing tundra. A huge pulse of methane would increase warming even faster, potentially leading to the nightmare scenario of rapidly melting ice sheets in Greenland and West Antarctica.

"If the Arctic starts to go, what are our options?" Caldeira asked during the first night's dinner. Everyone agreed that the only way to have a quick impact was to change the reflectivity of the region, which would cut the amount of heat absorbed from the sun. There was a brief mention of dumping white Styrofoam balls into the ocean and other half-nutty ideas. But in the end, the consensus was that there were really only two possibilities: either brighten ocean clouds to reflect sunlight or fill the stratosphere over the region with particles. In both cases, we could in theory cool the Arctic quickly enough to stop the ice melt more or less immediately.

As a way of shading the planet, cloud brightening has a lot of promise. But it turns out that the method works best as a way to enhance marine stratocumulus clouds, which are rare in the Arctic. There might be ways of getting around this, such as brightening clouds over the northeast Atlantic to cool the currents that flow up to the Arctic, but this is a tricky bank shot. There is also the problem of building the cloud-brightening machines themselves, which no one has actually done, and which would probably take a decade or so to perfect.

That left injecting particles into the stratosphere above the Arctic as the default option — what Alan Robock, an atmospheric scientist at Rutgers University who was at the dinner, called "the yarmulke solution." There was much discussion about what kind of particles would be best, how big they needed to be, how high they needed to be placed, how best to get them up there, how long they would stay in place, and what impact they would have. On all this, there was much disagreement. But there was no disagreement on the fact that it could be done quickly and — ignoring for a moment all the political, moral, and environmental implications — that it would cool the Arctic in very short order.

I'm not sure what it felt like to be in the room when NASA scientists finally decided that, yes, it was technically possible to put a man on the moon. But this was probably the closest I'll ever come to that experience. Ironically, it was Wood who best captured the scope of the discussion that night in San Francisco. "This is like nothing that human beings have thought about before," he said, his big hand wrapped around a Diet Coke, his manner uncharacteristically grave. "We are in uncharted terrain here."

When I visited David Keith in Calgary in 2006, two big questions on my mind were, What kind of person dreams of engineering the entire planet? and Can we trust him? I was asking myself those same questions the first time I met Lowell Wood. Shortly after my trip to Calgary to visit Keith, I heard about Wood's interest in geoengineering from a number of scientists, several of whom jokingly referred to him as Dr. Evil. I read a few of his papers on the subject, some of which had been coauthored with Teller, then called his office and asked if we could meet. I presumed he would decline. He was still working at Livermore at the time, and given his shadowy reputation and involvement in top-secret research, I presumed he would not be media-friendly. I was wrong. All it took was one phone call, and I was on my way out to California.

Despite Livermore's near-mythological status as the epicenter of high-tech weapons research, from the outside it looks like any other industrial facility—a collection of low, brown buildings surrounded by strip malls and rolling brown hills. Only when you get close do you notice the very sturdy-looking chainlink fence mounted with cameras every few hundred feet. Security at the lab has always been tight, but after September 11, 2001, local papers reported that the lab had acquired a truck-mounted Gatling gun, lest anyone be foolish enough to try to storm the gates and go after the thousand or so pounds of plutonium that is presumably stored on the grounds. Entry gates were guarded by Department of Energy troops wearing bulletproof vests and big pistols strapped to their hips.

As I rolled past the guards, I was reminded again of the strange pairing of guns and geoengineering. Climate and weather engineering has long had a military dimension, even apart from Teller's efforts. During World War II, the Germans experimented with creating fog to confuse Allied bombers. During the Vietnam War, the U.S. military used weather modification techniques, such as seeding clouds with silver iodide, in a secret attempt to increase rainfall over portions of the Ho Chi Minh Trail. None of these efforts worked very well, but that hasn't stopped the military from dreaming about such techniques.

Nor has it stopped the spinning of conspiracy theories. In fact, if you type the word "chemtrails" into a search engine, you'll find thousands of links to the websites of people who believe that the U.S. government has long been spraying metallic particles out of airplanes in a massive campaign to—well, nobody is exactly sure *what* the objective is. Biowarfare? Population control? Planet cooling? Whatever the purpose, chemmies believe that evidence of this stealth campaign can be seen in the numerous contrails of jets, which often linger in the air in suspicious and threatening patterns. According to chemmies, millions of people are now being poisoned by the barium and aluminum in these aerosols as they invisibly rain down on our heads. Naturally, one of the key figures in this conspiracy is (supposedly) a scientist at the Lawrence Livermore Lab.

Inside the gates, the lab is an oddly peaceful place. It feels like the grounds of a community college, except with brightly marked radiation shelters every few hundred feet. I parked my rental car in the designated area and headed to a small building where I had arranged to meet Wood. I expected to be greeted by a PR person or lab administrator who would oversee our conversation. When I finally arrived at the designated conference room, however, it was empty. I waited ten, fifteen minutes, wondering whether Wood had changed his mind about the meeting. Then I noticed a big, red-bearded man struggling with the lock on the side door. He was dressed in blue pants and a blue shirt, which was half-untucked. He was breathing heavily, sweat beading on his forehead. I thought he was a janitor who had come to clean the room. Only when I looked at his badge did I realize that he was the great weaponeer.

Wood said that he had no official title at the lab. "I'm just a physicist, an ordinary physicist," he told me. Actually, at first glance, he seemed more like a caricature of a mad scientist: sloppily dressed, disorganized, chatty and temperamental, his pockets stuffed with bits of paper with important notes jotted on them. In the 1960s, I later learned, when colleagues hid a lead brick in Wood's briefcase, he toted it around for days before noticing it. They told stories about him picking locks on Jetways when he was late for his flights and chartering private planes to rush him to nuclear test sites. And despite the fact that Wood has spent most of his adult life considering the best way to kill (or protect, depending on your point of view) millions of people, his emails to me were dotted with emoticons.

"Threats are my business," Wood told me. "I help the government figure out who can kill us, and how, and when." Talking with Wood, as you might imagine, feels a little like stepping into the middle of a Tom Clancy novel. When it comes to threats, Wood said, his top concern was an engineered pandemic, such as anthrax or smallpox. (If invited, Wood will go on a tear about the fact that the genetic blueprint for the smallpox virus has been published in academic journals. "Talk about an open invitation.") For kicks, I asked him about other threats—like, say, North Korea or Iran lobbing a missile at the United States. He shook his big head, amused. "A North Korean strike? What does it consist of? A Taepodong, if they can ever get it flying, with a nominal three- to ten-kiloton warhead on it. Give me a break." He's equally dismissive of the Iranians, who, he believes, buy everything from the Chinese. "These aren't threats," he tells me. "These are annoyances."

Wood's evolution into a Cold Warrior began when he was a student at UCLA in the 1960s. Teller was lecturing in physics at the university and was always on the prowl for smart young students whom he could draft to build nuclear weapons that would protect American freedom. Before long, Wood had completed his Ph.D. in astrophysics and moved to Livermore to work with Teller on miniature hydrogen bombs and nuclear fission. In their partnership, Teller was the big-picture guy, Wood the detail man. "Lowell was always a better engineer than Teller," physicist Freeman Dyson told me. "Teller loved big ideas but was not so interested in how to actually implement them." Both men became symbols of what President Dwight D. Eisenhower had called the military-industrial complex and were reviled by antiwar activists. In 1970, Wood was indicted for nuclear war crimes by a Berkeley group called the Red Power Family. Teller was burned in effigy.

In the 1980s, Teller and Wood developed the idea of space-based x-ray laser weapons that could vaporize Soviet missiles before they reached the United States. It was the ultimate technocratic hallucination, an umbrella of perfect defense created by launching a fleet of giant satellites armed with laser-producing nuclear bombs. Despite the project's technological complexities and huge costs, Teller managed to sell it to President Ronald Reagan, who was eager to fund anything that might rattle "the evil empire." The project, officially dubbed the Strategic Defense Initiative, but known to everyone as "Star Wars," became the centerpiece of the Reagan administration's defense policy. Billions of dollars in research money flowed to Livermore, with most of it going to support "O Group," which was charged with actually building the x-ray laser.

By most accounts, O Group was Wood's baby, a ragtag bunch of Berkeley and Stanford grads in their twenties who worked insane hours, fueled mostly by soft drinks and ice cream, driven by a sense of mission and pride in the fact that they were the smartest weapons builders on the planet. Richard Gabriel, a software industry pioneer who worked at the lab in the early 1980s, recalls that one of the guys had two pictures above his desk—a geographic map of the Soviet Union and a geographic map of the moon, labeled "before" and "after." According to Gabriel, some of the scientists would carry weapons in their cars to protect them from KGB agents and terrorists. In their rare free time, the group would hang out at the enormous log home Wood had hand-built in the hills above the lab, where they'd goof off in geeky ways, such as shooting rabbits with .357s or cranking open a gas line that ran through the property and lighting it, creating a thirty-foot-high tower of fire.

But the x-ray laser turned out to be a debacle, done in by engineering problems, cost overruns, and the rapid decline and fall of the Soviet Union. All in all, $60 billion was spent on research for the project—with nothing to show for it. In the late 1980s, Wood began championing a far cheaper, far less technologically sophisticated, non-nuclear space-based missile defense system, which he dubbed Brilliant Pebbles. By the time President Bill Clinton took office in 1993, however, the Soviet threat had been replaced by worries about rogue dictators and suitcase nukes. Brilliant Pebbles was out-of-date before the first prototype was built.

When the Soviet Union collapsed, Wood felt vindicated. "I was proud of the small part I played in this historic victory," he told me.

But what was a Cold War scientist to do now that the evil empire had been vanquished? Yes, the threat of a pandemic was interesting, but Wood was an astrophysicist, not a biologist. He liked doing battle in the cosmos, not behind the walls of immune cells. Then global warming began to emerge as a major issue. Now *that* was a planetary-scale threat worth thinking about. At the time of our first meeting in 2006, Wood wasn't entirely convinced that man-made pollution was the central factor in heating up the planet. He has now changed his mind on that. But even then, Wood admitted that, whatever was behind it, the consequences of a rapidly warming planet were pretty serious. Serious enough that we—a rich, civilized, technologically sophisticated generation of *Homo sapiens*—"might want to do something about it."

"Like what?" I asked him.

Wood shifted in his seat, then ran through a list of options—from launching big mirrors into space to dumping iron into the oceans—before dismissing them all as too expensive, too complicated, or just plain ineffective. "When we decide to solve this problem," he told me that day in Livermore, "we are likely to do it by doping the stratosphere."

In 1990, around the time Lowell Wood was putting the finishing touches on Brilliant Pebbles, Caldeira was packing up for a five-month trip to St. Petersburg as part of a scientific exchange program with the Soviet Union. He was still a grad student in atmospheric science at New York University at the time, living in downtown Manhattan, letting his hair grow down to his shoulders, suspicious of capitalism ("I was a quasi-socialist," Caldeira told me), and playing bass guitar in a post-punk band called Fist of Facts. Caldeira had taken a circuitous route into science, majoring in philosophy as an undergrad at Rutgers University, then working for a few years as a computer programmer at a big Wall Street firm before leaving to write software to track insider trading on the New York Stock Exchange. But even before he finished his Ph.D. at NYU, Caldeira was a standout: his first scientific paper, which challenged one of James Lovelock's hypotheses about how the oceans and clouds work together to modulate the earth's temperature, had already been published in *Nature*, one of the world's most prestigious scientific journals.

In St. Petersburg, Caldeira worked at the State Hydrological Institute, one of the leading research centers in the country. The top climate scientist at the institute—and in all of the Soviet Union—was Mikhail Budyko, who, at the age of seventy, was nearing the end of his career. Thirty years earlier, Budyko had revolutionized climate science by developing the first model of the earth's energy balance—calculating the flow of energy in (in the form of sunlight) and out (in the form of radiant heat). He had also been among the first to recognize that humans were indeed heating up the planet with greenhouse gas emissions. More to the point, his model had led him to think about how the earth might be cooled by changing its reflectivity, or albedo. One of the simplest ways to do that, Budyko suggested in the early 1960s, would be to mimic volcanoes, which inject large amounts of sulfur dioxide into the atmosphere. The gas eventually coalesces into tiny particles, which reflect sunlight. In theory, to build an artificial volcano that would have a similar effect, all you'd need to do is burn a big pile of sulfur (there's plenty of it around, and it's cheap) and funnel the sulfur dioxide gas up into the sky.

Caldeira didn't spend a lot of time talking with Budyko about this idea. At the time, it seemed to Caldeira to be part of a larger Soviet interest in meddling with nature, and Budyko was not exactly the kind of scientist who relished conversation with long-haired Americans. "I remember him as a hunched and hulking old man, with thin gray hair, with a serious demeanor and an air of formality," Caldeira said. Instead, Caldeira spent his time pursuing a woman named Lilian, a former researcher at the institute who would later become his wife, and working on the problem of thermal inertia in the climate system. He was attempting to calculate how much of the additional warming from rising CO_2 levels had already been expressed (only about half, he concluded—the rest of the warming had been absorbed by the oceans and had not yet radiated out).

When Caldeira returned to the United States, he finished his dissertation, which was a study of the earth's climate system at the time the dinosaurs went extinct, then spent a few years as a postdoc at Pennsylvania State University. In 1995, he took a job as a scientist at the Lawrence Livermore Lab, where he worked on a new climate and carbon model being developed by the lab. He focused on the impact of CO_2 on the oceans and within a few years produced several landmark studies about the damaging effects of high CO_2 levels on ocean chemistry. In fact, Caldeira coined the term "ocean acidification," which was used as the subtitle of one of his papers.

Livermore, of course, was the kingdom of Teller and Wood. The 1992 National Academy of Sciences report on geoengineering, the first rigorous examination of geoengineering options by some of the nation's top scientists, had intrigued Teller and Wood. It was in the NAS report that Wood first learned about the virtues of injecting particles into the stratosphere, which the authors pointed out was both cheap and (in theory) simple to do. The report concluded, "Perhaps one of the surprises of this analysis is the relatively low costs at which some of the engineering options might be implemented."

Teller and Wood started thinking about the engineering details of a particle injection scheme: How big should the particles be? What should they be made of? They decided that particles about one-tenth of a micron in diameter (a micron is about the size of the smallest dust mote visible to the human eye) would be most effective at scattering sunlight. These particles could be engineered out of some nonreactive metallic substance, such as aluminum, or better, generated from sulfates, substances readily available as byproducts of oil refining. And because the particles would be injected high into the stratosphere, Teller and Wood realized it wouldn't take much to have a big impact. The equivalent of one artificial volcano every five or six years would do the job.

By focusing hard on the engineering details, Teller and Wood calculated that you could cool the planet with aerosols for about $1 billion a year—far less than what Keith had estimated in his 1992 paper (see chapter 2) and several orders of magnitude less than the frequently cited costs of cutting CO_2 emissions. Like the atomic bomb, which Wood and Teller viewed as a device to keep evil in check, they saw injecting particles into the stratosphere as a technological solution to the problem of our self-destructive appetite for fossil fuels. What was not to like?

Plenty, as it turned out. When Wood first presented his ideas to a roomful of climate scientists at a 1998 conference at the Aspen Global Change Institute in Colorado, he hardly got a rousing welcome. Keith told me, "I was stunned to see Lowell Wood—this famous Star Warrior—there talking about global warming." Like other scientists in the room, Keith knew of Teller's past schemes to blast canals and harbors with nuclear bombs. Now his protégé was proposing to geoengineer the climate?

Nobody was more surprised by Wood's presentation than Caldeira. It was hard to imagine two men with more different backgrounds. In 1982, at about the time Wood was engineering x-ray lasers to blast Soviet missiles out of the sky, Caldeira was helping to organize a historic antinuke rally in New York's Central Park that drew nearly a million people. It was an odd coincidence that both he and Wood worked at Livermore now—although the two had never met. Caldeira was well aware of the enormous ethical and environmental complexities of Wood's proposal, which his presentation skipped right over. At one point, recalled Caldeira, Wood even joked that the best way to stop global warming would be to start a nuclear war. "It was pretty outrageous," Caldeira told me. "But now I realize it was just Lowell playing provocateur."

After Wood's talk, both Caldeira and Keith were very skeptical about the effectiveness of shooting dust into the stratosphere. In addition to having a general distrust of easy technological fixes promoted by infamous weaponeers, Caldeira, as a climate modeler, knew that heat from sunlight is different from heat trapped in the atmosphere by a thicker blanket of CO_2. Caldeira believed that one of the primary side effects of Wood's scheme would be to even out the temperature differences between night and day, to reduce seasonal changes, and to equalize the spread of heat between the equator and the poles. In other words, it would push the entire planet toward a more uniform and homogeneous climate.

In conversation afterward, Wood insisted that Caldeira was wrong. Caldeira replied that he was not wrong and that he would prove it using a state-of-the-art climate model running on the lab's supercomputers. It took him several months to run the simulation, but he was startled by the results: the particles quite effectively canceled the warming influence of CO_2, both regionally and seasonally. Wood's calculations, it turned out, were correct.

Caldeira's paper, which he coauthored with Bala Govindasamy, another climate modeler at Livermore, turned out to be a big boost for the legitimacy of stratospheric aerosols. It was one of the first solid modeling studies suggesting that, although there were still many unknown risks and side effects, a high-CO_2 world with geoengineering might be more like our own than a high-CO_2 world without geoengineering. In the paper, Caldeira stated unequivocally that the first priority should be to reduce emissions. But he concluded, "Our results suggest that geoengineering may be a promising strategy for counteracting climate change." Today many climate modelers question various aspects of Caldeira's study, but no one has refuted his central point: pumping particles into the stratosphere might be a viable way to reduce the risk of global warming.

After Teller's death in 2003, Wood kept refining and optimizing his ideas for how to inject particles into the stratosphere. But unlike Caldeira, who always bracketed these discussions with caveats and qualifications, Wood was all about engineering and advocacy. Indeed, he argued that shooting dust into the sky would have benefits beyond cooling the planet. Higher levels of CO_2 and more diffuse sunlight would increase agricultural productivity and create more colorful sunsets. ("Who doesn't love pretty sunsets?" he quipped.) The particles, he said, would also reduce damaging ultra-violet light, which causes skin cancer and kills nearly half a million people each year around the world, including sixty-six thousand in the United States alone. But most important, Wood kept stressing, was the price. If you wanted a cheap fix for global warming, the cli-mate equivalent of superglue, this was it.

Of course, Wood is no fool. In my conversations with him, he was careful to note that geoengineering is a "suboptimal" solution to global warming. But in his view, the moral, ethical, and environ-mental complexities of shooting particles into the sky are far out-weighed by the fact that we live in a culture that expects fast, easy solutions to whatever ails us.

A few months after my visit to the lab, Wood told me via email:

When I talk with people who object to geoengineering, I often say, "You don't have to argue with me, and I don't have to argue with you, let's find something more pleasant to talk about, because I'm going to win." It's just written in the stars. Geoengineering is go-ing to win, because the politicians, when they finally come down to the crunch, are going to ask: What is the cheapest thing that might possibly do the job? They don't care what it is; if it consists of Las Vegas dancers performing in the rotunda of the capital, they'll choose that if it's the cheapest solution. That's the way things work in a democracy. People never pay more than they have to.

Or as he put it in another email:

We serious geoengineers will just continue to wait oh-so-patiently until the political elites finally decide that [1] they've really got to stop temporizing-&-posturing, and instead do SOMETHING about increasing atmospheric CO_2, and [2] their individual-&-collective political lives indeed are at immediate risk, if they do ANYthing at all costly to ANY major constituency. At that point, they'll note that the Rio Framework Convention commands the

most cost-efficient remediation approaches—and they'll swiftly-
&-reliably beat a path to the Geoengineering Door. :-) History
(well, geophysics-&-economics) is on Our Side! :-)

After their encounter in Aspen, Wood and Caldeira became friends.
But it was an odd friendship, based not on shared political ideology—Caldeira jokingly refers to Wood as "a right-wing nut," while
Wood calls Caldeira "a tree-hugger of the most liberal persua-
sion"—but on a mutual respect for bold ideas and a belief in the
virtue of free thinking. They meet once in a while for lunch or dinner near the Stanford campus, bantering about the news of the day
and, in an informal way, teasing out their latest thoughts about how
we might cool the planet.

At one such lunch in the spring of 2006, Caldeira and Wood
were joined by their friend Gregory Benford, a noted science fiction
writer and physics professor at the University of California, Irvine.
They were discussing the latest data on summer sea ice in the Arctic,
when Benford said, "Why don't we do an experiment in the Arctic
and see if we can cool it off?" Benford himself is an outspoken advocate of geoengineering and believes that solving global warming is
not a technical or engineering problem, but a moral one. "The standard American puritan view of the environment is that doing less
is virtuous," Benford told me shortly after the lunch. "In the case of
global warming, doing less is catastrophic."

Benford's enthusiasm aside, Caldeira and Wood agreed that a
real-life experiment in the Arctic was, for now, out of the question.
But that didn't mean they couldn't run simulations. After some discussion, Caldeira and Wood decided to crank out some modeling runs to see if shooting particles into the stratosphere over the
North Pole could help stabilize the region. Exactly how much sunlight would have to be reflected to stop the ice from melting? What
effect would it have on the rest of the earth's climate?

Caldeira plugged some numbers into the climate model at Stanford, and after several weeks of complex calculations, he had some

answers. Shading the sunlight by 20 percent above 70 degrees latitude—essentially the region directly over the polar ice cap, where no humans live—would maintain the "natural" level of ice in the Arctic, even with a doubling of atmospheric CO_2 levels. Push it up to 50 percent, and the ice would grow. Even better, the restoration of the ice would happen fast, resulting from a drop in temperature of more than 3 degrees Fahrenheit in five years.

The news of the modeling results interested Wood, who began calculating exactly what it would take to shade the sunlight by 20 percent. (This is just the kind of cosmic calculation that Wood loves to noodle around with.) He believed that about 200,000 metric tons of particles each year would do it. On a planetary scale, that's a tiny amount—and only a small percentage of what we dump into the lower atmosphere every year by burning fossil fuels. As for how to get those particles up into the stratosphere, Wood calculated that a fleet of modified 747s or high-flying military aircraft could do the job. Or a twenty-five-kilometer Kevlar tube, with a diameter slightly larger than a garden hose, could be built to pump the particle-producing gas up there. The bottom of the hose would be connected to a combustor that would create the gas, while the top would be held in place by a high-tech kite or the kind of high-altitude airship that the U.S. Department of Defense is currently developing for other research purposes ("It's nothing more than a fancy blimp," Wood said).

Wood's analysis overlooked a lot of important details, such as how to keep the particles from sticking together once they were sprayed up into the sky. But like his mentor, Edward Teller, he was more interested in big ideas than small details. To him, this plan to save the Arctic was drop-dead simple. It would stabilize the ice, save the polar bears, and demonstrate the virtues of planetary engineering for less money than it takes to feed and clothe our soldiers in Iraq for a year. Because the aerosols are launched only over the Arctic, there is little fear of directly impacting humans. And best of all, the system could be tested for a few years to see whether it worked.

If something went wrong, the test could be aborted, and within a year or so, all the particles would have dissipated, returning the region to its "natural" state.

In the summer of 2006, a few months after he and Caldeira finished their model runs, Wood presented his plan to save the Arctic to an elite group of scientists, economists, and government officials gathered at the Snowmass ski resort near Aspen, Colorado. The weeklong workshop, held in the shadow of fourteen-thousand-foot peaks at the Top of the Village lodge, was organized by the Energy Modeling Forum, an independent group of academics and industry leaders affiliated with Stanford University.

After Wood ran through his PowerPoint presentation, reactions came fast and furious. A number of attendees, including Richard Tol, an economist with the Economic and Social Research Institute in Ireland, found Wood's ideas innovative and worthy of further research. A number of others, however, were outraged by the blithe, unscientific, speculative, downright arrogant proposal of this . . . *weaponeer*. After all, it was one thing to write a speculative scientific essay about geoengineering, as Paul Crutzen had done, but it was quite another to work out ways to actually do the job. Shooting particles into the sky might cause unexpected droughts, one scientist at the meeting argued, or shift the monsoons, or monkey with ocean circulation patterns. Who the hell knew what might happen. Bill Nordhaus, a Yale economist, worried about political implications: wasn't this simply a way of enabling more fossil fuel use, like giving methadone to a heroin addict? If people believed there was a solution to global warming that did not require hard choices, how could political leaders ever make the case that people needed to change their lives and cut emissions? Stanford professor John Weyant, surprised by the "emotional and religious" debate over Wood's proposal, cut off discussion before it turned into a shouting match.

Wood was delighted by the ruckus. "Yes, there was some spirited discussion," he boasted a few days later as he sipped a Diet Coke in a Mexican restaurant in Silicon Valley. "But a surprising number of people said to me, 'Why haven't we heard about this before? Why aren't we doing this?'"

Then Wood flashed me a devilish grin. "I think a few of them were ready to cross over to the dark side."

When it comes to shooting particles into the sky, few scientists doubt two basic premises of Wood's argument: it would be cheap to do (in comparison to other geoengineering schemes, as well as greenhouse gas reductions), and it would cool the earth quickly. But beyond that, things get more complicated.

It may not be technologically complex to loft particles into the stratosphere, but there is no system in place to do it. In one recent study, aerospace experts went beyond Wood's preliminary engineering designs and looked more deeply at the practicality of particle injection. They explored some of the more far-out ideas that Wood had suggested, such as using hoses tethered to high-altitude balloons, but discovered that the quickest and least expensive way to do it would be with high-altitude aircraft. The trick is finding planes that can carry a big enough payload and still reach the altitude necessary (30,000 to 40,000 feet) for the most effective particle dispersal. One of the best planes for the job, the study found, might be an experimental aircraft called the White Knight Two, a strange-looking plane with three carbon-fiber fuselages that is designed to fly as high as 60,000 feet, taking the plane's pilots and passengers right to the edge of deep blue space. (British entrepreneur Richard Branson plans to use White Knight Twos in his private space flight company, Virgin Galactic.) The study found that a fleet of 150 planes, making two flights a day, would be required to lift the necessary mass of particles up to the stratosphere. The total cost to cool the planet, including planes, materials, and fuel, would be $8 billion a year. That's about eight times more than Wood had estimated, but in the grand scheme of things, it's peanuts.

Another delivery method, recently floated by people with knowledge of high-tech military hardware, is to use hydrogen guns, which are an ultra-high-tech version of a spring-piston air gun and can fire projectiles very long distances at high velocity. (At the first dinner in San Francisco during the 2008 AGU meeting, Wood offered still another idea that was even faster and easier: "Make a pile of the material you want to loft into the stratosphere. Then set off a nuclear bomb in the middle of it. That will get it up there." I'm fairly sure he was joking.)

Another practical issue is exactly how to create and then disperse the particles. Spraying a sulfur gas into the air is the most "natural" method — it essentially mimics a volcano — but building a system that will create particles of optimal size and then keep them from sticking together and raining down from the sky is not trivial. David Keith and a colleague have proposed using hot sulfuric acid vapor to create the particles, which might give researchers better control over particle size and distribution. (You wouldn't want to mess around with hot sulfuric acid vapor in your backyard, but up in the stratosphere, it's nothing to be alarmed about.) Wood likes the idea of engineering particles out of an inert metallic substance, such as magnesium or aluminum, which would allow the size and design to be optimized for maximum reflectivity, thereby greatly reducing the amount of material needed to do the job and making the task of lifting it into the stratosphere that much easier. Keith has also come up with an idea for engineering particles that loft themselves into the atmosphere using a process called photophoretic levitation. It's a complex and fantastical concept, based on the fact that small particles tend to migrate when exposed to intense beams of light. But if it worked, it could someday eliminate the need for jets or Kevlar hoses. In theory, these particles could also be controlled from the ground, allowing the world's climate czar — whoever that may turn out to be — to have control over what would amount to a louvered shade for the planet.

As for environmental consequences, one emerging concern is the impact that particles in the stratosphere would have on ozone chemistry. "The chemistry of the earth's atmosphere is exceedingly complex," Michael Oppenheimer, a professor of geosciences and international affairs at Princeton University, told me. Oppenheimer has pointed out that the injection of particles could trigger the creation of additional chlorine in the stratosphere, the chemical most damaging to the ozone. Indeed, at least one study has found a significant impact on ozone levels, especially in the polar regions. "From my point of view," Oppenheimer said, "you're trading one destructive environmental problem for another—not a good idea either in the short run or the long run." But other scientists have argued that natural particle-spewing events such as the eruption of Mount Pinatubo in 1991 destroy less than 3 percent of the world's ozone layer, which suggests that the damaging effects of particles might be small. "Ozone loss needs to be looked at carefully," Caldeira said. "But right now, it doesn't look like a showstopper."

A much bigger problem is the effect that higher CO_2 levels will have on the oceans, even in a world that is artificially "cooled" by a man-made sunshade. Oceans act as big sponges, soaking up atmospheric CO_2, which forms carbonic acid when dissolved in water. For years, climate scientists assumed that the earth's natural buffering capacity prevented changes in acidity even with massive increases in CO_2 levels. But in 2003, Caldeira and a former colleague at Livermore, Michael Wickett, calculated that the absorption of high levels of CO_2 could make the oceans more acidic over the next few centuries than they have been for 300 million years. What will be the effects of increased ocean acidification? No one knows for sure, but coral reefs are likely to dissolve, as will the tiny plankton whose shells or skeletons are made from calcium carbonate and which are a key link in the ocean food chain. So unless shooting particles into the stratosphere is combined with sharp cuts in emissions, it may well save the climate but destroy the oceans.

"There are an enormous number of complexities here, an enormous number of relationships that we're just beginning to understand," said Stephen Schneider, a climate scientist at Stanford University and one of the early critics of geoengineering. "The only context in which it makes sense to talk about this is if we're in a true planetary emergency."

Perhaps the most hotly debated issue, however, is the effect that particles in the stratosphere will have on the monsoons. The Asian monsoon, which sweeps through Indonesia, India, Thailand, and China, is the main source of water for large areas of the most heavily populated continent. An estimated two billion people—nearly one in three people on the planet—rely on it to grow their food. Take away the monsoon, and they will starve. Ditto with the African monsoon. Climate scientists know that reducing sunlight, whether it is by throwing particles into the stratosphere or any other technique, will also reduce evaporation—and all else being equal, that will reduce rainfall. But by how much? And will reduced evaporation be offset by changes in atmospheric dynamics, which could bring more moist air over land? (Monsoons are driven largely by temperature differentials between land and oceans.) Getting trustworthy results from climate models for specific regions is notoriously difficult, especially in the tropics. And even if precipitation declines, that does not necessarily mean that there will be a parallel decline in food production. (Any decline in precipitation could be countered by increased CO_2 from rising emissions and more diffuse sunlight, both of which should make plants more productive.) On top of everything else, before you can make a judgment about the consequences of changes in the monsoon patterns, you have to take into account the bigger picture. And as Caldeira often points out, all credible computer models have shown that the climate of a geoengineered world with reduced sunlight would be more like today's climate than one in which CO_2 levels are doubled and nothing is done.

Finally, there is the Sword of Damocles problem, which I touched on in chapter 1. Particles in the stratosphere last only a year or so at best, so in order to keep the earth cool, the supply would have to be constantly replenished. If the supply was stopped, it would be like closing a parasol on a hot day—we'd suddenly feel the full force of the sun. So if we used particles over a few decades to mask a warming of, say, 5 degrees Fahrenheit, as soon as we stopped injecting them, the planet would heat up by 5 degrees virtually overnight. Talk about climate shock! And because CO_2 lingers in the atmosphere for thousands of years, it is likely that this Sword of Damocles would hang over our heads for a very long time.

There may be ways around these dilemmas. Tom Wigley, a climate modeler at the National Center for Atmospheric Research in Boulder, Colorado, has suggested that if spraying particles in the stratosphere were combined with an aggressive push to reduce greenhouse gas emissions, the aerosols could "take the edge off" the current warming trend and buy us more time to develop carbon-free energy supplies. (The more diffuse sunlight created by particles in the stratosphere would have no impact on photovoltaic solar cells, but it would lessen the efficiency of large-scale solar thermal installations, which rely on bright, direct sunlight.) If CO_2 were quickly reduced at the same time that we pumped particles into the sky, it could potentially stabilize the ice melt in the Arctic, reduce the danger of ocean acidification, and, in the long run, make it less dangerous to taper off particle injections—although to avoid an abrupt jolt to the climate system, we would still need to keep injecting particles for decades, if not centuries. As for the Sword of Damocles, we have lots of swords hanging over our heads already, from the generation of electricity to the pumping of water to the production of nitrogen fertilizer. Stopping any one of these things immediately would be catastrophic—and yet we rarely think about it that way.

All of these uncertainties, of course, are reasons why many climate scientists advocate more research into the costs and consequences of geoengineering. "The more we know, the better we will be at gauging risk," Keith told me. And climate modelers like Caldeira caution against thinking of particles as an all-or-nothing approach. "We could start with something very modest, maybe one one-hundredth of full deployment each year, which may be less

risky than just letting greenhouse gas concentrations rise," Caldeira explained. "If you start out gradually and you see something bad
start to happen, you could taper off. The gentler you are with the
climate system, the less trouble you are going to get in. If you hit it
hard, you are more likely to cause a problem."

But the real wildcard, in Caldeira's view, is how we humans
would respond to a geoengineered world. Putting particles in the
atmosphere is likely to cause sunlight to become more diffuse, blur-
ring the sharp lines of shadows and turning the sky whiter during
the day, while making sunsets redder in the evening. How do you
calculate the psychological impact of that? It's not just the end of
blue skies as a physical reality but also the end of blue skies as a metaphor — that suggestion of the great beauty and power of nature,
of a promising day ahead, of cosmic beneficence. Instead of a sky
above us, it would be a ceiling. Would this man-made artifact put a
roof on our aspirations and dreams? Would it make us feel as if we
were living in a cage? Or would it be something that would seem
strange for a few years but that everyone would get used to?

Wood has heard all these arguments before and isn't much impressed. After all, when you're used to thinking about the world in
terms of a nuclear Armageddon, why worry about a little change
in sky color? During the dinner meetings in San Francisco, it was
clear that Wood believed that the question of whether we will or
won't end up spraying particles into the stratosphere has little to do
with environmental consequences or engineering complexities. It
will be decided by our dysfunctional political system and our love
of a quick fix. For Wood, it's just a matter of time until the world
comes around to the inevitability of geoengineering. As he put it in
an email to me, "The future is ours, Comrades."

SEVEN

....

A Little Cash on the Side

ONE OF THE MOST frequently cited objections to geoengineering is that the earth's climate is just too complex for us to mess around with. "We're fiddling with a very complicated system, trying to counter the consequences of other large human influences, inadvertent ones, in that system," said John Holdren, President Obama's science adviser. "And it's a dicey business, because we're doing it without a complete understanding of how the machinery works."

Obviously, there is much truth to this, especially given the potential consequences of a mistake. (*Oops, sorry, didn't mean to turn France into a desert!*) But it is also true that messing with things we don't understand is what human beings do. Isaac Newton, confounded by the mysteries of white light, messed around with a prism and invented the modern science of optics. Thomas Edison messed around with electric current; today we have light bulbs, laptops, and a digital economy. Curiosity and exploration are what drive scientific discovery, and until the twentieth century, when scientists started fiddling with genetic structures and nanomaterials and atomic bombs, no one was too concerned that an aggressively inquisitive scientist was going to wreck the planet. Newton may have poisoned his brain by trying to smelt lead into gold, but that was his problem.

In the world of geoengineering, mistakes — even if they're well-intentioned — are everyone's problem. Holdren is exactly right when he talks about geoengineering being dicey business — but it only becomes dicey when you move from talking about it and thinking about it to actually doing it. In this sense, it's a little like robbing a bank. It's perfectly fine to sit home all day and draw up plans; it's only when you walk up to the teller and actually pull out a gun that you're asking for trouble.

Consider the notion that we can remove CO_2 from the atmosphere by dumping iron into the oceans. As an idea, it's pretty interesting. Just as we have deserts on land, we also have deserts in the sea — not because of a lack of water, of course, but because of a lack of nutrients (especially iron) that allow plants to grow. If you dump a few hundred tons of iron into the barren waters near Antarctica, it's like pouring Miracle-Gro on your garden: suddenly, phytoplankton (the scientific term for the single-cell plants that live in the surface waters) are blooming everywhere. Like all plants, phytoplankton suck up CO_2. In fact, phytoplankton are ruthlessly efficient little carbon-eating machines. Although they make up less than 1 percent of the photosynthetic life on earth, they are responsible for about half the CO_2 that is pulled out of the atmosphere each year (the other half is pulled out by land plants). The phytoplankton use the carbon to build calcium carbonate shells. When the organisms die, they slowly drift down into the deep ocean, where the carbon is effectively trapped for hundreds to thousands of years. Some of them settle on the ocean floor, where their shells are eventually transformed into carbonate rocks, such as limestone, emerging millions of years later in spectacular outcroppings like the white cliffs of Dover, England, or becoming the chalk that my daughter Grace uses to draw happy faces on the sidewalk in the spring.

There is very little mystery in any of this. The life and death of phytoplankton are part of a natural carbon cycle known as the biological pump. This cycle has been operating since phytoplankton began to evolve about 1.5 billion years ago, and if we don't cook the planet, it's likely to be happening 1.5 billion years hence.

That is not to say there aren't many unresolved questions about this process, especially if we're considering dumping iron into the ocean as a solution to global warming. Among them: How much CO_2 can we actually remove from the atmosphere this way, and is it enough to make a difference? How do we measure exactly how much additional CO_2 is being sequestered by a given ton of iron that is dumped into the ocean? (A good percentage of the artificial plankton bloom is likely to be eaten by other creatures, thus working its way up the food chain and ending up who knows where.) Then there are questions about the environmental impact: What effect will the changing nutrient balance in the ocean have on other sea life? Will it lead to the ocean equivalent of a runaway weed patch? What will it do to fish and mammal populations in the region?

Answering these questions is not beyond the pale of scientific inquiry. And in a rational world, they would all be resolved with a reasonable degree of certainty before people took to the high seas with boatloads of iron to save the planet. But we don't live in a rational world. We live in a world that is populated by ambitious entrepreneurs who want to make money and, as an added bonus, stop global warming. And that's where the trouble begins.

A few years ago, I spent a month in the North Atlantic on the *Knorr*, a 267-foot research vessel operated by the Woods Hole Oceanographic Institution in Massachusetts. The purpose of the trip was to study the changing nature of the ocean by taking mud cores from the sea floor several hundred miles off the East Coast, at the edge of the continental shelf. The mud, which is hundreds of feet deep in this area, contains the ocean equivalent of dinosaur bones: the shells of single-cell planktonic organisms called foraminifera, or forams, which live in the sea by the billions and whose shells perpetually rain down onto the ocean floor. The chief scientist on the trip, Lloyd Keigwin, a paleoclimatologist at Woods Hole, was interested in forams because a number of species build their shells in rhythm with the temperature of the oceans they live in, making their discarded shells an excellent thermometer for past climates. With the help of sophisticated tools such as a mass spectrometer, Keigwin can measure the chemical and atomic structure of the old shells and use this information to create a vivid picture of the conditions in the North Atlantic, say, 8,500 years ago.

Before my trip on the *Knorr*, I knew nothing about any of this. When I thought of ocean life, I thought of whales and sharks and dolphins. On the *Knorr*, I learned to see the ocean in a different way. Each evening, we set out a small dragnet—a kind of ultrafine sieve—behind the ship and pulled out a sample. Then we examined our catch under one of the powerful microscopes in the onboard laboratory. It was like peering into life on another planet. The planktonic universe was alive with bizarre creatures at war with one another, spiral-shaped organisms drifting around, twiggy animals swimming and devouring all in their path, and, most of all, an infinite variety of phytoplankton—strange blooms more various than snowflakes, as beautiful as spring flowers in my wife's garden.

The purpose of this cruise had nothing to do with iron fertilization. I was onboard simply to learn more about how scientists reconstruct past climates and what that can tell us about our future. But I ended up learning a lot about plankton, both dead and alive, and about how the vigor and diversity of this planktonic world is closely connected to the chemistry of the ocean, just as the vigor and diversity of the ecosystems on land are closely connected with the chemistry of the soil. On the *Knorr*, we passed through the luminous blue waters of the Sargasso Sea, which many oceanographers refer to as a marine desert, because it's devoid of nutrients commonly found in other parts of the ocean. The Sargasso is dominated by a unique type of seaweed that rolls like tumbleweeds along the surface and is particularly well adapted to the region. But beyond that, the Sargasso has little plant life (which is partly why the waters are so blue — it is the phytoplankton that give water its greenish hue). Why are certain regions of the ocean so desolate? Until recently, most oceanographers believed that the abundance of planktonic life rose or fell depending on the level of just two nutrients: nitrogen and phosphorus.

John Martin, a well-known oceanographer who headed Moss Landing Marine Laboratories in California from the late 1970s to the early 1990s, believed it was more complicated than that. Martin, a brilliant biogeochemist whose ability to do field research was handicapped by the effects of polio, suspected that an absence of trace metals might explain these desolate zones in the ocean. Martin eventually hypothesized that iron functions as a micronutrient and that the absence of iron limits phytoplankton production. A simple experiment with two flasks of ocean water from the Antarctic proved him correct: when he added iron to one sample, the phytoplankton bloomed.

The implications of this discovery were huge. Martin, among others, was able to establish a clear connection between iron levels and CO_2 in the earth's climate history. Here's how it works. Iron, Martin knew, is carried into the ocean by dust storms that blow off the continents. (Soil contains trace amounts of iron.) As the climate cools and dries out, dust storms increase, which fertilizes the oceans with iron, stimulating phytoplankton blooms. These blooms draw more CO_2 out of the atmosphere, which further cools the planet, causing more dust storms, and the loop continues. Martin discovered that iron is one of the control levers on the biological pump. Although he would never have described himself as a geoengineer, he immediately understood the implications of his discovery as a tool for climate control. At a science conference at Woods Hole shortly before his death in 1993, he joked, "Give me half a tanker of iron, and I'll give you an ice age." In the years since, that line has been quoted in nearly every media report about ocean fertilization. The fact that it's obvious hyperbole is rarely noted.

Still, any technique or tool for drawing down the level of CO_2 in the atmosphere is clearly of interest to geoengineers. Although removing CO_2 is not going to have an immediate impact on the earth's temperature in the same way that blocking sunlight with particles would, it is potentially an important tool for long-term climate management. In this sense, iron fertilization is not unlike David Keith's air-capture machine or, for that matter, a tree farm. Given how dire the climate crisis has become, it's hard to imagine any kind of global climate management plan that would not involve pulling CO_2 out of the atmosphere. One approach does not preclude the other — removing CO_2 and blocking sunlight are just different tools to fix the same problem. In fact, the most likely scenario is some kind of ever-changing mix of technologies based on regional conditions and the shifting responses of the climate system.

Pulling CO_2 out of the atmosphere has one big advantage over blocking sunlight, however: you can make money doing it. There is no market for shade, or for reducing the watts per square meter of sunlight hitting the earth. There is, however, a market for carbon — and it's huge. The Kyoto Protocol laid the groundwork for a global carbon market back in 1997, envisioning a future in which tons of CO_2 are swapped like pork bellies or barrels of oil. The European Union launched the first large-scale trading scheme in 2005, and by 2008 it was worth $120 billion. By some estimates, the global carbon market could reach $3 trillion by 2020.

Despite the battering of Wall Street's reputation during the financial meltdown of 2008–2009, it's pretty clear that carbon trading will be the mechanism of choice in most developed countries for dealing with greenhouse gas emissions. To some, this might seem like asking for trouble — David Brower, the legendary head of the Sierra Club, once called the idea of using markets to address environmental problems "a form of brain damage" — but at this point it is close to inevitable, if for no other reason than that it is the only greenhouse gas reduction scheme that Wall Street knows how to make a buck on. In addition, the U.S. Acid Rain Program, which uses a cap-and-trade system to reduce sulfur dioxide emissions from coal-fired power plants, has been too successful to ignore.

The basic idea is simple: instead of passing laws that mandate a particular kind of technology to reduce pollution — such as catalytic converters on cars or scrubbers on power plants — a "cap," or annual emissions quota, is set for each polluter. (Caps can be applied to individual power plants, to companies, or to entire nations — the principle is the same.) Polluters that reduce emissions below their caps can sell the rest of their allowances — that's the market term for permits to emit greenhouse gases — to other companies or bank them for future use. Polluters decide for themselves the best way to meet their caps: install a pollution control device like a scrubber, switch to a cleaner fuel, and/or buy allowances from other polluters. The point of the program is to encourage polluters to discover the most creative, low-cost ways to cut emissions. And, of course, smart players figure out that they can make money by reducing emissions below their caps and then selling the balance of their allowances.

In the case of the Acid Rain Program, this strategy has worked beautifully. The program has cut sulfur dioxide emissions in half since 1990, and has done it more cheaply than anyone thought possible. The program has also saved lives. The Environmental Protection Agency estimates that by 2010, annual human health benefits — mostly in reduced lung disease and its related effects — will exceed $50 billion. The *Economist* called the sulfur dioxide trading program "the greatest green success story" of the 1990s.

It was the success of the Acid Rain Program, in fact, that led to emissions trading being incorporated into the Kyoto Protocol. Like the Acid Rain Program, the trading program agreed to in Kyoto (and expanded in subsequent revisions to the treaty in Bonn and Marrakesh in 2001) requires that each developed nation adopt a cap on greenhouse gas emissions, then allows those nations to meet their caps either by reducing greenhouse gas levels or by purchasing allowances from other countries that have reduced their emissions below their caps.

But there are important distinctions between the market for sulfur dioxide emissions and the emerging carbon market. Greenhouse gases are very different from pollutants such as sulfur dioxide, most of the impacts of which are regional. In contrast, no matter where a molecule of CO_2 is emitted, the effect on the atmosphere is the same. Under a section of the Kyoto agreement called the Clean Development Mechanism, developed countries can get carbon credits — an alternative to allowances, but with essentially the same function — for investing in projects in developing countries that offset greenhouse gas emissions. In such countries, the cost of making reductions is often much lower than in developed countries. For example, a Danish power company might earn carbon credits by setting up a system to capture methane from a rotting landfill in Brazil which would otherwise be released into the atmosphere, or by erecting wind turbines in China that offset dirty coal power.

Exactly what constitutes a legitimate and verifiable offset is a matter of much argument and debate. Should polluters be able to earn credit for investing in, say, no-till farming, which helps keep carbon stored in the soil? Or for planting trees? What about for stimulating plankton blooms in the ocean? As it is structured today, the Kyoto agreement doesn't allow trading of carbon credits involving carbon sinks, such as afforestation or ocean fertilization. But given how much pressure there is to include offsets — polluters like them because they increase the supply of carbon credits on the market, thus lowering the price — the rules could very well change in the future. And if they do, it will start a stampede of entrepreneurs setting up new projects to pull CO_2 out of the atmosphere.

There are many ways to suck up CO_2, and each has its own strengths and weaknesses. Planting trees is the simplest method, with the added virtue that it stabilizes soil and creates new habitats for animals. Scientists and carbon-market regulators have given a lot of thought to forestry projects and have a decent understanding of how much carbon they sequester over what timescales. But planting trees is slow, labor-intensive, and more or less useless as a response to a climate emergency. Building air-capture machines like Keith's might be a better way to get CO_2 out of the atmosphere quickly, and it's fairly easy to calculate how much CO_2 is removed. But the technology is expensive and a long way from being ready for prime time.

As I mentioned earlier, James Lovelock is enthusiastic about the potential of biochar, which sequesters CO_2 by burying charcoal (which, of course, contains the carbon that was sucked out of the air through photosynthesis when the tree or plant was alive). Biochar has a number of important side benefits, including increasing the fertility of soils, and the amount of carbon sequestered is fairly easy to monitor. But it requires an infrastructure of kilns and stoves to burn biomass and create the charcoal. And it's not clear whether biochar technology can scale up quickly enough to have a big impact—although Lovelock believes that, if we got every farmer in the world to cook and bury his or her crop residue, we could sequester hundreds of millions of tons of CO_2 each year.

The virtue of another option, iron fertilization, is that it's cheap, quick, and very easy to scale up. It requires no more sophisticated technology than a big boat and a tank full of iron. Actually, you don't even need a tank full of iron. The most efficient way to fertilize the ocean is to dissolve the same substance that many golf courses use to keep the grass green—ferrous sulfate monohydrate, commonly known as iron sulfate—in seawater onboard the ship, then dribble the mixture overboard with a large hose. By some estimates, stimulating a plankton bloom costs up to a hundred times less than planting trees for an equivalent amount of CO_2 removed. There is no waiting patiently, year after year, for leaves to unfurl and roots to grow. If you pick the right spot, you can dump the iron slurry into the ocean and see the plankton bloom almost immediately. Yes, there are lots of questions about environmental impacts on the ocean ecosystem, as well as large uncertainties about how much of the CO_2 taken up by the phytoplankton in the surface waters will not reenter the atmosphere for more than a hundred years. But the outsize potential for moneymaking means that many of these questions get flattened by stampeding entrepreneurs.

To illustrate the potential, let's do the math. In today's carbon market, which currently exists only in Europe but is widely assumed will kick into gear in the United States in the near future, each ton of CO_2 that is not emitted is worth, at the time I'm writing this, about $25. The cost of dumping iron into the ocean is now about $5 to $7 per ton and, with practice, could fall to as low as $2.50 per ton. A single boat, working a patch of ocean for ten days or so, can dump four hundred tons or so of iron sulfate into the ocean. If conditions are right, that might stimulate enough phytoplankton to fix about three million tons of CO_2 (roughly the amount that a modest-size coal power plant emits in a single year). Of that, maybe one million tons might actually be sequestered in the deep ocean. That means that in ten days, a single boat could earn carbon credits worth $25 million.

Of course, it's not quite that simple. But almost. And even if credits for ocean fertilization are never approved for the global carbon markets, there is always the so-called voluntary market—that is, companies sell carbon credits to people who are eager to offset their carbon emissions by paying to reduce CO_2 pollution elsewhere. The value of a ton of CO_2 is not as high in voluntary markets, but the rules are often looser. So instead of $25 million for a ten-day iron dump, you might earn only $5 million.

Because so much money is at stake, ocean fertilization has become a test case for a lot of the difficult questions about geoengineering, none of which is more important than this: at what point does the urgent and heroic goal of fixing the planet become just another excuse to make a quick buck?

As it happened, I first met Russ George about three hours after I met Lowell Wood. I had set up back-to-back appointments, racing from the Lawrence Livermore Lab over the hill to the shabby Silicon Valley offices of the Planktos Foundation, which George had founded a few years earlier. Unlike Wood, who has a broad, speculative intelligence and playful (if eccentric) manner, George came across as a man on a mission to save the planet. He was in his mid-fifties, with long silvery hair and a trimmed beard, dressed in a blue Hawaiian shirt and flip-flops. He had founded Planktos, he told me, "because the earth is in deep trouble. The climate is heating up, and the oceans are dying. We can't go on this way, or it will be the end of life as we know it." His goal, he said, was "to advance research of iron fertilization as a way to restore plankton forests in the oceans, increase fish stocks, and reduce CO_2 levels in the atmosphere. We can feed the world, restore the oceans, and cool the planet all at once."

Okay, so he was a bit blustery — far more blustery, I couldn't help thinking, than Wood, who actually had something to be blustery about (the guy did build nuclear bombs and x-ray lasers). Ten minutes after I met him, it was clear to me that disaster porn was George's stock-in-trade. His business plan was to sell himself as a superhero.

On the Planktos website, George described himself, accurately, as "a lifelong ecologist wanderer and environmental professional with a range of accomplishments." But like Wood, George grew up in the shadow of Cold War science. His father was a nuclear physicist who worked on the Manhattan Project. "There were always Geiger counters around when I was a kid," he told me. George, however, was not interested in Geiger counters and nuclear physics — at least not at first. He studied biology in college, then founded a tree-planting company called Coast Range in British Columbia in 1972. He lived in a tent with his hippie friends, bidding on forest restoration projects funded by the timber industry. Later, he drifted through other environmental jobs, including a stint, as he later claimed, "standing night watch at sea on the Greenpeace ship *Rainbow Warrior*."

When his hippie days ended, George returned to the atomic world that his father helped pioneer. In 1992, he founded a company to research and commercialize a technology that a handful of scientists believed was just slightly less important than the discovery of fire: cold fusion. Instead of splitting atoms, as in nuclear fission, cold fusion joins them. A similar reaction powers the sun, although that is, of course, hot fusion. In theory, cold fusion releases similarly huge quantities of energy. Even better, the reaction can take place in a simple apparatus the size of a postage stamp and does not emit radiation. The trouble is, most of the claims for cold fusion are based on one highly controversial experiment in 1989 — which no one has been able to replicate, leading many nuclear physicists to believe the whole notion is junk science. But not George. "Probably the most difficult hurdle is that cold fusion seems too fantastic scientifically and too good to be true economically and socially," George said in a 1992 interview. "But cold fusion will likely revolutionize the world in ways we can barely begin to imagine."

Luckily for George, other business opportunities emerged. In 1997, the Kyoto Protocol was ratified, creating an instant market for carbon credits. Although forestry offsets were not included in the initial agreement, it was a good bet that they might be included in future amendments. Smelling opportunity, George began what he called "Canada's first major eco-restoration climate forest company," called HaidaClimate, which was a partnership with the native people of the Queen Charlotte Islands. He also started a partnership with the government of Hungary, where he proposed planting more than 100,000 acres of new forests on degraded farmland. As the managing director of the company put it bluntly, "We believe the quality, affordability, and sustainable volume of the credits we will be offering are truly unprecedented in the carbon trading world."

Tree planting, however, is not the ideal medium for an ambitious entrepreneur — it's slow, hard to scale up, and, with all the complex land-use regulations, bureaucratic. In contrast, creating what George called "a plankton forest" in the ocean was practically magical. Just mix up an iron sulfate slurry, spill it out the back of a boat, and voilà! Instant bloom. Instant carbon sequestration. Instant cash. And as long as it was done in international waters, no pesky bureaucracies were likely to interfere.

By 2002, George was living on a boat moored in Half Moon Bay, just over the hill from Silicon Valley, and dreaming of two things: cold fusion and plankton forests. Of the two, plankton forests were the safer and more profitable bet. He started the Planktos Foundation (to call it an organization may be too strong a word — it was mostly just him in those early days) to give himself a tax-free way to raise and spend money on ocean fertilization research, with an eye toward turning it into a profitable business as soon as possible. He didn't have the cash to buy an oceangoing research vessel from which to conduct his studies, but he did talk the captain of Neil Young's hundred-year-old wooden schooner *Ragland* into letting him borrow it for an expedition to Hawaii. During the trip, George dumped a few tons of iron into the sea, made a few measurements of the phytoplankton bloom, and declared the whole adventure a success.

George was not the only one clued in on the economic potential of ocean fertilization. One company, Greensea Venture, recruited leading oceanographers to join its mission and proposed dumping iron in an eight-thousand-square-kilometer area of the Pacific Ocean. Another company, Carboncorp USA, also promoted sequestration through fertilization, describing a process by which it might dump a proprietary fertilizer mix from commercial ships that routinely cross the high seas. The Ocean Technology Group at the University of Sydney patented an "ocean nourishment" process in which ammonia would be piped into coastal waters to stimulate plankton blooms. Kenneth Coale, director of Moss Landing Marine Laboratories, told one reporter that a Japanese company hoping to fertilize the oceans for carbon credits offered him $1 million if he would lend his name to its operation.

You can't get into the ocean fertilization business without a boat, however (Neil Young's generosity notwithstanding). For that, George needed a financial backer. In 2005, a friend introduced him to Nelson Skalbania, a well-known sports and real estate mogul in Vancouver. Skalbania, who owned a fleet of Rolls-Royces, was a flamboyant character who had been convicted for stealing $100,000 from a partner in a real estate deal in 1997 and narrowly escaped going to jail. He was well-known for investing in do-good projects that employed fringe science, including a company that planned to use energy from water evaporation to create electricity. "He is the kind of guy that you go to when you have some crazy idea and a sliver or a hair chance of getting it financed," Wolfgang Richter, a Victoria developer, told Canada's *Globe and Mail.* Skalbania was taken by George's idea for ocean fertilization, eventually investing $4 million. "I thought it might do some good for the world," Skalbania told me.

George also tried to recruit Dan Whaley, a young Silicon Valley entrepreneur, to invest in Planktos and take over as CEO. Whaley, who was in his early thirties, was one of the golden boys of the dotcom boom in the Valley. In 1995, he cofounded GetThere, one of the first travel booking sites on the Internet. In 1999, Whaley and his partners sold the company for $750 million. *Salon,* an online magazine in the Bay Area, voted him one of the top ten bachelors in Silicon Valley. After his big score, Whaley spent a few years remodeling his Moroccan-style house on a lagoon in San Mateo, installing a ten-person stone hot tub and other luxuries, and spending as many daylight hours as possible wakeboarding. Eventually, he decided it was time to do something more meaningful with his life. He took a soul-searching drive through South America, where he read books about peak oil and global warming and began to think seriously about the future of human civilization. Shortly thereafter, Whaley first heard about Russ George and his plan to fertilize the oceans.

A friend arranged an introduction, and before long Whaley invited George to his house for a barbecue. They spent a few hours talking about George's ideas. Whaley was intrigued. For George, who desperately needed a smart rich guy like Whaley to run his operation, Whaley was an ideal partner. At George's invitation, Whaley spent a few months at the Planktos office, learning more about the science and business of ocean fertilization. It just so happened that Whaley's mother, Margaret Leinen, was assistant director for geosciences at the National Science Foundation and very well connected with ocean scientists.

Ultimately, Whaley decided not to invest in Planktos — "I liked Russ, but he was just too brash and unpredictable, and he had weird, shifty eyes" — but he did see the virtues and economic potential of ocean fertilization. Shortly after Whaley parted ways with George in 2006, he decided to start a rival company, which he called Climos. Unlike George, Whaley was able to line up some first-class investors — a list that eventually included the Contrarian Group, the investment company run by former Major League Baseball commissioner Peter Ueberroth, as well as Elon Musk, a cofounder of PayPal and Tesla Motors, an electric car company. George wasn't happy about Whaley's move — he later threatened to sue Whaley for stealing company secrets — but it certainly validated George's idea that he was onto something important.

By the end of 2006, the Planktos Foundation had officially become the Planktos Corporation, and George was convinced that he had bootstrapped a big idea into a world-changing business. "The oceans are dying," he told me. "The planet is getting hotter. If we don't fix it, we're headed for catastrophe."

During our conversations, George never mentioned the word "geoengineering." What he was engaged in, he said, was a "restoration project." His pitch, I learned, was an artful mix of science and spin. He talked about his deep respect for John Martin and his discovery that iron is a critical micronutrient for plant growth and photosynthesis. Unfortunately, George argued, ocean plankton have been decimated in recent years because their usual windborne iron dust supply had dwindled by a third. Planktos would fix that by giving back what nature had taken away ("replenishing the shortfall") — a sort of Geritol for the oceans. And the impact would be enormous. "Full restoration" of the open ocean's plankton to 1980 levels, he said, would cancel out roughly half of all manmade CO_2 emissions each year. Dumping iron into the ocean would not only revive the plankton, but it would buffer ocean acidity, cool the sky, and "recharge the entire marine food chain." And that, in turn, would help feed starving people throughout the world by increasing fish stocks. All in all, it was a pretty impressive pitch for his product, one that reduced all the world's problems down to a single boatload of iron.

George told me about his plans for an upcoming research trip, which was scheduled for the summer of 2007. Somewhere out in the Atlantic—he wouldn't say where—the Planktos crew planned to fertilize a sixty-square-mile section of ocean with about fifty tons of iron, releasing it from the back of the boat while traveling back and forth in rows. "It's like plowing a field," George said. He expected that fifty tons of iron would sequester three million to five million tons of CO_2 from the surface waters.

In my conversations with George, it was clear that he believed that getting rich and saving the planet were not contradictory goals. He mentioned more than once that financial analysts believed that ocean fertilization could be a $100 billion business in the not-too-distant future. And although George admitted that many scientific questions about iron fertilization were far from settled, Planktos was already selling "Greentags"—basically, certificates for unverified CO_2 reductions resulting from the company's ocean experiments that he marketed to people who felt guilty about their carbon footprint—for $5 a ton. In addition, George made sure that I was aware that he had been on the phone that very day with executives from American Electric Power, the largest coal-burning utility in America, who were looking for a cheap way to offset emissions. And he noted that power consumption at a recent electric utilities conference in Arizona had been entirely "carbon neutral" thanks to offsets provided by Planktos.

"My motto," George volunteered, "is 'Save the world and make a little cash on the side.'"

The emergence of start-up companies such as Planktos set off alarm bells in the scientific community. It wasn't that anyone doubted

Martin's iron hypothesis—a number of experiments had shown that in iron-limited areas, particularly the Southern Ocean around Antarctica and the northern Pacific, adding iron did indeed stimulate plankton blooms. And nobody was too concerned that a few small-scale experiments would destroy the ocean. What spooked scientists was the lack of regulatory oversight or governance of potential iron entrepreneurs. It was not hard to imagine, in the not-too-distant future, a flotilla of boats suddenly appearing off the coast of Brazil, each dumping many tons of iron into the sea and looking to cash in on saving the planet. There was still a lot that scientists didn't understand about how ocean fertilization worked and what the consequences would be—environmentally and politically—of a sudden rush to cultivate the oceans.

It's not that scientists hadn't spoken out about it. In 2001, a year before George made his first research expedition to Hawaii, Sallie Chisholm, a highly respected ocean researcher at MIT, was the lead author of a paper called "Dis-crediting Ocean Fertilization," published in the prestigious journal *Science*. The paper was a passionate takedown of entrepreneurs such as George who claimed that "ocean fertilization is an easily controlled, verifiable process that mimics nature; and that it is an environmentally benign, long-term solution to atmospheric CO_2 accumulation." These claims, the authors wrote, "are, quite simply, not true."

Among other things, they pointed out that dumping iron in the sea from ships does not replicate the dust storms that nature creates. Those storms gradually add iron to a region over long timescales. The impacts of a sudden large infusion of iron on ocean chemistry and, more important, the subtle links in the food chain were impossible to calculate. Some models predicted that iron fertilization could result in a shortage of oxygen in the deep ocean, which in turn, the authors wrote, could "shift the microbial community toward organisms that create greenhouse gases such as methane and nitrous oxide, with much higher warming potentials than CO_2." In other words, instead of cooling the planet, iron fertilization could inadvertently help heat it up.

Another major uncertainty was what happens to the CO_2 after it is consumed by the phytoplankton. How much actually makes it to the deeper waters, where it is sequestered for hundreds of years (or more)? How could it be measured and verified? And how could we be certain that when the plankton bloom was increased in one area, it wouldn't use up nutrients that would otherwise have been available for phytoplankton blooms in other areas? Sorting this out, the authors suggested, would "require complex numerical models of large-scale ocean physics and biogeochemistry."

Finally, there was the larger cost-benefit analysis of what commercial-scale ocean fertilization would do to the planet: "To us, the known consequences and uncertainties already far outweigh hypothetical benefits," the authors wrote. And even in a hypothetical best-case scenario, only about 15 percent of all CO_2 that humans release into the atmosphere would be sequestered through fertilization. Was it really worth it? Weren't there better, safer ways to reduce CO_2 levels—like, say, shutting down a few dozen big coal plants?

In the view of Chisholm and her coauthors, the solution was not to stop researching ocean fertilization, which produced valuable science, but to remove the profit incentive: "We suggest that ocean fertilization, in the open seas or territorial waters, should never become eligible for carbon credits."

This paper, of course, had zero impact on George's plans. Nor did subsequent papers suggesting that only between 5 and 50 percent of the CO_2 that plankton blooms pull out of the surface waters actually makes it down into the deep ocean. By 2007, climate modelers such as Ken Caldeira were estimating that even in a best-case scenario, iron fertilization wouldn't draw down anything like 15 percent of all CO_2 emissions, as Chisholm and her coauthors suggested. It was more like 4 percent.

Of course, this did not mean that iron fertilization was not worth thinking about. Given the urgency of global warming, any idea or technology that could draw down CO_2 was worth exploring, if only to understand exactly how useful—or how damaging—it might be. If anything, it made research into ocean fertilization more important than ever. And small-scale fertilization experiments could potentially reveal a lot about the complex geochemistry of the oceans. But George was not really interested in complex geochemistry or an honest cost-benefit analysis. He was interested in getting his boat in the water.

By early 2007, Planktos had invested $1 million in its own research vessel, a 110-foot steel workhorse called the *Weatherbird II,* and Planktos shares were a hot commodity on penny stock exchanges. George's ideas might have been half-baked, but his timing was impeccable. He was lauded by the media as a wacky-but-maybe-brilliant entrepreneur who played into the romantic fantasy of the lone scientist who saves the world. Few reporters asked hard questions about what he was really up to or what the larger implications of his crusade might be. Instead, they giggled at images of Pico, a smiling cartoon of a phytoplankton, which Planktos's PR literature described as "our wild and crazy mascot, our coccolithophorid sidekick."

In reality, George's emergence on the public stage was troubled from the start. In a move that suggested just how tone-deaf he was to the challenges Planktos faced, the company announced that it would soon begin a field test near the Galápagos Islands, where the ocean conditions, George said, were particularly suitable for iron fertilization. It apparently didn't occur to him, or to David Kubiak, Planktos's PR guru, that the idea of dumping iron into one of the most fragile and celebrated ecosystems on the planet might ignite some controversy. (Kubiak, a founding member of 911Truth .org — a conspiracy group that believed the U.S. government was not telling the full truth about the terrorist attacks in 2001 — clearly has his own ideas about how the world works.) Planktos tried to spin the Galápagos destination to their advantage — press releases called the field test "a voyage of discovery," playing off Charles Darwin's famous trip to the islands more than a century earlier — but that only underscored the flimsiness of the company's own scientific credentials.

In the early spring of 2007, George courted the media with a press junket to Washington, D.C., complete with tours of the *Weatherbird II* (he had sailed it up the Potomac) and a news conference at the National Press Club. "I cannot overstate the importance of these Planktos pilot projects," said former United Nations Environment Programme director and Planktos chief scientific adviser Dr. Noel Brown. "If their applied science works half as well as the early research indicates, this work will both help restore the neglected oceans and give everyone concerned about global warming truly meaningful hope."

Nice thought, but Brown was an administrator, not a scientist—a distinction that was surely lost on many at the press conference. (His Ph.D. is in international law and relations.) The company continued to exaggerate the potential for ocean fertilization, while downplaying any risk or environmental consequences. Planktos's website boasted that the technology "offers investors the single most powerful, profitable, and planet-friendly tool in the worldwide battle against global warming." True or not, the hype seemed to be working. Although the company's revenue amounted to only a few thousand dollars, raised by selling tons of credits to individuals and small businesses largely through its online "store," George's little entrepreneurial start-up had a market value of more than $90 million.

One unintended consequence of Planktos's PR campaign was that it alerted environmental activists to the company's plans. For them, any kind of geoengineering is anathema, and Planktos was an irresistible target. On May 2, 2007, as the crew of the *Weatherbird II* were gathering supplies for their trip across the Pacific, a small coalition of environmental groups, including Greenpeace and the Ecuadorian nonprofit Acción Ecológica, issued a press release under the heading "Geoengineers to Foul Galápagos Seas." Planktos's plans were a "risky gamble with sensitive marine ecosystems," the statement read. "Climate change is a real threat, but common sense should not be its first victim." As Brent Blackwelder, the president of Friends of the Earth, later put it, "Planktos is selling the equivalent of global warming snake oil."

George was not happy, especially when one environmental group seized on Planktos's online description of its iron particles as "nanosize," warning that "the Planktos experiment may be the largest intentional release of engineered nanoparticles ever undertaken." In fairness to George, that statement grossly misrepresented what Planktos was up to. In George's view, calling iron sulfate "engineered nanoparticles" was not only dishonest but also a calculated attempt to use heavily loaded language to raise unfounded fears in the public. As for trashing the Galápagos, George countered that local environmentalists knew that prevailing currents would carry any iron away from the islands, not toward them. In his view, the crew of the *Weatherbird II* were like organic gardeners, helping the ocean restore and enrich itself in a way that was safe and natural.

Had George not been chasing a buck, his arguments might have carried some weight. As it was, he just fit the stereotype of a profiteer who was willing to play fast and loose with science. The Sea Shepherd Conservation Society, a Greenpeace splinter group of self-professed "eco-pirates" committed to stopping whaling by any means necessary, took notice of Planktos's plans. The group called the plans "a dangerous experiment" and vowed to stop them. Sea Shepherds regularly patrol the Galápagos, where they confront illegal long-liners, sea-cucumber poachers, and other outlaws. From 1979 to 2007, the group had rammed and sunk nine ships and, just as important, proved itself to be enormously skilled at courting positive media coverage. Captain Paul Watson, head of the organization, called Planktos "a scam to make money" and made none-too-subtle threats about the Sea Shepherds' plans for a high-seas encounter with the *Weatherbird:* "We are not Greenpeace. We won't be just showing up to hang banners and take snapshots."

The most crippling blow for Planktos, however, was financial. In September 2007, the International Environmental Law Project wrote a letter to the Securities and Exchange Commission, which was signed by environmental groups, arguing that Planktos's appeal to investors was "false and manipulative." The letter said that Planktos had made inaccurate statements about the applicability of U.S. laws to its work, implying that there were no environmental regulations that could interfere with the company's ability to operate a commercial-scale ocean fertilization business. In fact, there were a number of laws and treaties that Planktos might run afoul of, including the London Convention, which regulates ocean dumping, even in international waters, from U.S. vessels unless they have a permit from the EPA. The notion that Planktos was playing games with the SEC underscored the shadiness of its operation, and before long the company's stock price plummeted.

By the time *Weatherbird II* finally cast off from its berth in Florida on November 5, 2007, more than three months behind schedule, the plan to dump iron near the Galápagos had been scuttled. Where were they going instead? Because of what Planktos called "the frontier nature of the research," the company kept the ship's destination secret. That just increased suspicion and made the ship's departure an international story, picked up by everyone from Reuters to Andrew Revkin, the respected environmental reporter for the *New York Times*. The story line was the same: what the hell are they up to? It was a real-time soap opera, tailor-made for the newly emerging blogosphere.

All this drama and intrigue, however, was lost on the dozen or so crew members on the ship, who had no access to the media. George himself was not onboard, nor was the ship staffed by geoengineers. In fact, Peter Willcox, the fifty-four-year-old captain of the *Weatherbird*, was the last person you'd expect to be piloting a ship on such a controversial mission. In a world full of wannabe eco-warriors, Willcox was the real thing. He had captained Greenpeace's *Rainbow Warrior* for nearly two decades and was in charge of the ship in 1985 when, shortly after returning to a New Zealand harbor after an antinuclear protest on a Pacific island, the ship was bombed and sunk by the French secret service, killing one crew member.

Willcox, of course, knew exactly where they were going: the Canary Islands, a Spanish territory 150 miles off the western coast of Morocco, where, he was told, the water chemistry was right, the currents were suitable, and Planktos had established a research partnership with a local university.

Willcox had never heard the word "geoengineering," and the notion that there was anything controversial about Planktos's mission never occurred to him. In fact, like many people, he liked the idea that entrepreneurs were getting involved in the fight against global warming. But Willcox never asked too many questions of George, accepting his scientific conclusions at face value. Once they put out to sea, however, it did strike Willcox as odd that there were no scientists onboard. "There were a few kids with marine biology degrees, but the experiments that I saw were all high-school-level stuff," Willcox told me. At one point, he and others jumped off the boat to swim with some whales. Someone onboard snapped a picture, and, according to Willcox, it later ended up on the Planktos website with the caption "Planktos Scientists at Work."

The *Weatherbird* stopped in Bermuda for a few days to refuel and restock, then cruised east across the Atlantic. By early December, the ship neared the Canary Islands. The plan was to take on more supplies—including a fifty-ton load of iron—and begin the first experiment in nearby seas.

When the ship was about twenty miles off the Canaries, Willcox radioed the harbormaster and asked for permission to enter—standard practice when approaching an international port. To his surprise, permission was denied.

In all his years at sea, this had never happened to Willcox before. He radioed back, explaining that they were on a scientific expedition and that Planktos had arranged for a berth in the harbor. More important, several crew members had recently become violently ill with stomach cramps and vomiting. "I have sick people onboard," Willcox explained. "They need to see a doctor."

Sorry.

Unbeknownst to him, environmental activists had convinced Spanish authorities that the *Weatherbird II* was carrying toxic waste that the crew intended to dump in Spanish waters. The charge was untrue, but as a tactic, it worked: Spanish authorities wanted nothing to do with the ship.

"We did donuts in the ocean for three days," Willcox recalled, then was told to head for Funchal, a harbor on the Portuguese archipelago of Madeira. George met the ship a few days later, and as the crew members recovered from their illness, there was an ongoing discussion about whether to continue the expedition. Now the whole enterprise was imploding: the environmental community was up in arms, the company stock was on the verge of being delisted, and, thanks in part to all the bad publicity surrounding the voyage, the chance that anyone would ever trust Planktos as a source of carbon credits seemed remote. Nelson Skalbania, who was bankrolling the operation, had had enough. "I decided to pull the plug," Skalbania told me. The crew was paid and cut loose, the *Weatherbird II* was sold off, and a company that just a few months earlier was valued (on paper) at more than $90 million vanished overnight. All told, Skalbania lost $4.5 million. Michael Bailey, a Planktos employee, called it "a perfect storm of corporate collapse."

In the coming months, George defended himself on the Internet, blaming the company's demise on "a highly effective disinformation campaign waged by anti-offset crusaders." But Willcox, for one, had a different interpretation. After spending some time boning up on the science of ocean fertilization, he realized that he had been duped by George's "save the world" rhetoric. "In retrospect," Willcox told me, "the whole thing looks like a Russ George get-rich-quick scheme."

If you believe that greed, properly directed, can be a force for good in the world, it's easy to blame the crackup of Planktos on George's problematic character and strategic blunders and let it go at that. If George had been a little savvier, as well as a little more open to working with scientists and environmentalists, he might well have been able to conduct his experiment without harassment. After all, not even Sallie Chisholm was arguing that dumping fifty tons of iron into the ocean, by itself, was going to do much damage. (It was the idea that Planktos would inspire a stampede of iron entrepreneurs that worried her.) And, if George had been a little shrewder, this argument continues, he might have both contributed to our scientific understanding of how iron can be used to draw CO_2 out of the atmosphere and taken a modest step toward building a profitable business in ocean fertilization. After all, good science is good science, no matter who pays for it or what the motives of the researcher may be.

This is essentially the argument that Dan Whaley, the CEO of Climos, made to me from a barstool in San Francisco a few months after the Planktos meltdown. The bar was in the South of Market district, the epicenter of the dot-com revolution in which Whaley had made his money. The buzz was gone now, but Whaley still looked the part of the young, ambitious entrepreneur — dressed in jeans and an open-neck shirt and exuding an air of informality and earnestness as he sipped his beer.

During our conversation, Whaley made it clear that, superficially, Climos might have had a lot in common with Planktos — after all, Whaley had gotten the idea for Climos after spending a few months poking around at Planktos — but, in fact, their strategies couldn't be more different. When it came to ocean fertilization, Planktos was the bad cop, Climos the good one. Climos had respected scientists on its board, it was working hard to build bridges with environmental groups, it had participated in all the right scientific conferences, and, most important, it had real money behind it. Whaley's mother's reputation as a serious player in the U.S. science establishment didn't hurt either. (Perhaps self-conscious of the fact that his chief science adviser is his mom, he refers to her as "Margaret" in conversation.) "We are not a pirate operation," Whaley said to me, half-joking. He talked about his own deep concerns about global warming, and about how important it is to leverage the creativity and know-how of private industry to solve the problem. I had no doubt that he was sincere.

If Whaley were just another cleantech entrepreneur, nobody would care if he was running a pirate operation or not. You build a cheaper, better solar panel, and you change the world. It's that simple. But when you're talking about dumping iron into the ocean — or shooting particles into the sky, for that matter — it is an entirely different matter. You are not just building hardware; you are messing with the entire operating system of the planet. Who decides how dangerous an experiment is? When there are strong laws and social norms that watch out for the public good, the risks of these kinds of experiments are manageable. But in the new world of geoengineering, where there are few norms and rules, the risks are much greater.

At least there *were* very few rules or norms until Planktos came along. One of the consequences of the Planktos debacle was to wake up regulators, watchdog agencies, and global bureaucrats to the problems of ocean fertilization and, more generally, to the trouble that could be caused by lone actors — imagine Russ George with a few hundred million dollars at his disposal — who are hell-bent on "restoring" the planet.

Not that there was much that could be done. A few months after Planktos flamed out, members of two major international treaties — the United Nations Convention on Biological Diversity and the London Convention — essentially agreed with the argument that Chisholm and her coauthors had put forth back in 2001: ocean fertilization must be banned as a commercial endeavor. But given the fact that the oceans are public domain, owned and policed by no one, passing a ban was easier said than done. As a first step, members of both conventions passed moratoriums on large-scale ocean fertilization schemes, while agreeing that legitimate small-scale experiments should be permitted. These moves sounded good, but in fact they led to much debate over the definition of "small-scale" and what constitutes legitimate scientific research. There was also the small matter of enforcement: there were no cops cruising the world's oceans, enforcing the ban with billy clubs. Nor was there any meaningful mechanism to punish countries that might decide to ignore the ban and allow ocean fertilization experiments to be carried out under their flags.

Ultimately, all this wrangling over the legal, ethical, and environmental aspects of iron fertilization killed Whaley's business plan for Climos. Whaley and his mother argued for softer language in the moratoriums, to no avail. Despite questions about enforcement and other issues, it was clear that commercial ocean fertilization was not going to be sanctioned by international agencies anytime soon. For Climos, this made it all the more difficult to cast itself as the voice of scientific integrity, and harder still for investors to see where the revenue stream for the company was going to come from. In 2008, Climos shifted its strategy and began investigating a range of other geoengineering technologies. "We're still interested in ocean fertilization, but we're no longer pursuing the carbon credit market," Whaley told me with some bitterness. Leinen cut her ties with Climos and started a nonprofit organization called the Climate Response Fund, "to stimulate and support discussion and research into geoengineering or climate intervention techniques."

On one level, you could argue that the moral of Planktos was that good science triumphed over bad. "The whole idea of allowing private companies to decide where and how much iron to dump in the ocean," David Keith told me with characteristic bluntness, "is nuts." Keith, along with Ken Caldeira and many others who have thought broadly and deeply about geoengineering, believe that in the long run, ocean fertilization may indeed have a role to play in pulling carbon out of the atmosphere. But it is likely to be on a relatively small scale, where the water chemistry and currents are right and the movement of carbon can be closely monitored. And if it's going to be done, they argue, it's probably best done not by private industry but by governments, where management of the public good is the top priority (at least in theory). As Keith put it in an email to me: "If we decide to dump iron in the ocean, I'm happy to let free enterprise figure out the most efficient way to do it. Just as if we decide to build tanks I'm happy to let free enterprise figure out the best way to supply them. But the idea that free enterprise should figure out how much iron to put in the ocean is as crazy as the idea that Lockheed Martin should figure out which country to invade."

On another level, good science was a loser in the Planktos story. If knowledge about the risks and dangers of geoengineering is going to be advanced, it is going to require more than just computer modeling. It is going to require field experiments. And the debacle of Planktos — and, to a lesser extent, Climos — will make it that much easier for anti-geoengineering activists to paint scientists as enemies of the environment. "Planktos and Climos created a tremendous number of antibodies against any field-testing whatsoever," Lowell Wood observed. Indeed, this is one reason why Caldeira and others fought hard to stop commercial operations like Planktos and Climos. "Ocean fertilization is just a headache," Caldeira told me in 2009. "Even in the best-case scenario, it's unlikely to help much — and it may even do more harm than good. And the controversy and hysteria surrounding it make it all the harder to test any ideas that might actually work."

....

EIGHT

The Romance of Clouds

UNTIL RECENTLY, I never gave clouds much thought. They floated in the background of my life, objects I glanced up at now and then, hoping they would bring rain or hoping they wouldn't, depending on my mood. Once in a while, I sat in the grass with my kids as they pointed out the alligators, ducks, and bears floating across the sky. In airplanes, I sometimes noticed towering castles of cumulonimbus or silver carpets of stratocumulus. But mostly, clouds were invisible to me, as they are to most people who live a largely urban life. Like stargazing, cloud spotting is an art of an earlier world, one practiced most often today by pilots, farmers, and curious children.

But you can't think much about global warming, much less about how to geoengineer the planet, without considering clouds. They are the most elusive and unpredictable aspect of the earth's climate system. Clouds can both heat and cool the planet, depending on their type and location. They are carriers of moisture, so understanding how they move is crucial to understanding rainfall patterns—and how those patterns might shift as greenhouse gas concentrations rise. But from a scientific point of view, the most remarkable thing about clouds is how infinitely variable they are. You can divide clouds into various types—most of us remember stratus, cumulus, and cirrus from elementary school—but these are only loose categories. In fact, every cloud is different—it has its own particular dynamics, its own particular chemistry, its own particular micro–weather system within it. This is part of what makes them so fascinating to the researchers who study them. It's also what makes them so complex for climate scientists and, more important, for potential geoengineers.

But thinking too much about the mechanics of clouds is like thinking too much about the mechanics of sex—it takes all the fun out of it. My own reawakening to the strange beauty of clouds began when I discovered an organization called the Cloud Appreciation Society. I stumbled across it the way most people stumble across things these days—during a Google search. The society, which is based in England and has more than eighteen thousand paying members, was started by Gavin Pretor-Pinney, editor of the *Idler,* a magazine that celebrates do-nothingness in all its forms. Cloud watching, of course, is about as unpuritanical an activity as one can imagine. Pretor-Pinney started the society, he says, because he felt that the wonders of cloud watching were being lost in our busy, head-in-a-cubicle digital world. Naturally, the society has a wonderful website and is in fact a great example of the fetishistic gatherings that are empowered by the Internet.

The Cloud Appreciation Society is exactly what it sounds like—a place to rediscover the wonders of clouds. "Nothing in nature rivals their variety and drama; nothing matches their sublime, ephemeral beauty," Pretor-Pinney wrote. On the website, you can participate in cloud chat with other cloud spotters or roll through thousands of images of clouds arranged by category. In one category, called "Clouds That Look Like Things," I found striking photos of clouds shaped like a locomotive, a lobster, and a tennis player. The website also has poetry written by cloud spotters. But my favorite feature is the society's manifesto, which includes this: "We believe that clouds are unjustly maligned and that life would be immeasurably poorer without them. We seek to remind people that clouds are expressions of the atmosphere's moods, and can be read like those of a person's conscience."

To say that clouds are the expressions of the atmosphere's moods is to personify nature in a way that makes scientists (except perhaps James Lovelock) cringe. But I found myself drawn to this kind of talk, perhaps because it is a reminder of how the world looked before we fully rationalized nature. Indeed, after I spent a few days trolling around on the society's website, I was eager to talk to my six-year-old daughter, Grace, about what she was seeing in the sky. As it turned out, one of the children's books she was reading at the time was set in seventeenth-century Japan and was about two kids coaxing rain out of a dragon in the sky. As I talked on the phone in my home office, I often found myself staring out the window, tracking the evolution of white puffs above the treetops. When I booked airline tickets, I chose a window seat instead of an aisle. I chuckled uncomfortably when a friend said, "Clouds are the face of God," but I understood what she meant. At a Bruce Springsteen concert, I was mesmerized by clouds racing across the big video screen behind the band during "Thunder Road."

And I began to really like the idea that we might cool off the planet by manipulating clouds. In the hard industrial world of geoengineering, playing with clouds is the only idea that has any romance. Not only does manipulating clouds seem safe and easy, but it actually seems fun — the geoengineering equivalent of fingerpainting in the sky. And although romance doesn't count for much when the planet is in crisis, it does count for something. Of course, it also has to work. And playing with clouds, I learned, is where elegant ideas bump into inelegant reality.

Stephen Salter, age seventy-one, handed my son Milo, age eleven, two small jars of glass balls and said, "This is going to save the world."

We were in Salter's office at the University of Edinburgh in Scotland. The office looked as if it had been decorated by Jules Verne.

It was stuffed with dusty books, wires, nuts, bolts, brackets, and other mechanical debris. Salter is widely regarded as an engineering genius of the nineteenth-century variety — a man who works not in digital bits and bytes, but in hard metals that obey the laws of physics. His most famous invention, known as Salter's Duck, is a wave energy machine that he built in the 1980s and that is still considered to be one of the most efficient devices to extract energy from waves (although it has never been deployed on a commercial scale — Salter blames the nuclear power industry in Britain for killing it). But he has also worked on robots, minesweepers, hovercraft, and fighter planes. Just down the hall from his office at the university is his metal shop, full of lathes and presses and other serious equipment. He prides himself on his ability to make things himself and happily repeats the story of how he was once dismissed by a white-shirted atomic engineer as "a greasy-fingered mechanic." ("I took that as an enormous compliment," Salter told me.)

Milo took the two jars from Salter. He is a smart kid, but he clearly had no idea what to make of Salter's statement about the glass balls saving the world — and frankly, neither did I. The two jars looked identical, except that the beads in one were darker than those in the other.

"How is that going to save the world?" I asked Salter.

He ignored me. "Take a close look at them," Salter said to Milo. He is a tall man with graying hair and a hawk nose. He was dressed in blue slacks and a blue Windbreaker. He has a reputation for being difficult to work with, a man excessively attached to the perfection of his own ideas, but with Milo, he was patient and interested. It occurred to me that designing cloud-brightening devices and wave energy machines is not unlike playing with Legos: to be a great inventor, you need to see the world in a childlike way.

"What's the difference between them?" Salter asked.

"The beads in one jar are darker than the other," Milo replied accurately.

"Yes — but why?"

"Because they are different colors."

"Ah, I thought you would say that. Actually, the beads are the same color. They are just a different size, so they reflect light differently. In the jar with the small beads, there are more beads, which have more surface area—so they scatter more light, which makes them look lighter."

Milo nodded, processing the idea. "Okay . . . So how is this going to save the world?"

"Because we're going to use the same principle to brighten clouds, which will then reflect more sunlight, which will then help to cool off the planet." Salter nodded to a chair next to his supremely messy desk, motioning for Milo to sit down. "Let me show you what I've been working on."

I don't usually have the luxury of taking my family with me on work trips, but I had brought Milo and his twin sister, Georgia, with me to Edinburgh because I had to travel to the city for a small workshop on cloud brightening and the workshop happened to be on their eleventh birthday. What better place to celebrate your eleventh birthday than in the shadow of Edinburgh Castle? My mother came along, too, both for the adventure and to help with the children while I attended the workshop. The day after the meeting, when I mentioned that I was going to visit the lab of a scientist who was working on a cloud-brightening machine, Milo immediately asked, "Can I come?" (Georgia preferred to explore the city with her grandmother.) I was reluctant to agree, mostly because I had not met Salter before and didn't know how he'd respond to having an inquisitive kid around.

But when we met outside his lab at the university, Salter seemed delighted that I'd brought Milo with me. He asked Milo about his interests, and when Milo mentioned science, Salter made it clear that he thought of himself as first and foremost an engineer: "Scientists are people who know more and more about less and less, while engineers have to know a little about a lot of things, and they have to learn it fast."

As we walked, I asked Salter about his background. He told us that he was born in South Africa in 1938 and lived there until he was seven years old, when his family moved to southeastern England. As a kid, he was always interested in making things — especially model airplanes. When he was seventeen, he began an old-fashioned apprenticeship with an aircraft company on the Isle of Wight. ("My grandmother was very concerned about social status. She said, 'You won't be with actual workmen, will you?' But of course I was. And I was proud of it.") He worked as a fitter, a toolmaker, and an instrumentation engineer. When he learned that some of the planes he was building were used for military purposes, however, he quit. He spent three years studying physics at Cambridge University, where he became interested in, among other things, artificial intelligence and robotics. In the early 1970s, he moved into mechanical engineering, especially wave energy, where his work is still considered revolutionary, even if it has never been commercially deployed.

In his office, Salter pulled up an artist's rendering of one of his cloud-brightening boats on his computer screen. He told Milo that each boat would be about 150 feet long and weigh 300 tons. It was a trimaran, with three long, canoe-shaped hulls linked together for stability. Instead of sails, however, there were three long tubes ("They looked like toilet paper rolls," Milo told me later) with regularly spaced ribs on them. These tubes, which would be about sixty feet high, were Flettner rotors, which act like sails but are more efficient. They would also work as smokestacks for tiny saltwater droplets that Salter planned to launch into the clouds.

I had seen the image of the vessel many times before — it's a favorite of scientists who talk about geoengineering, as well as journalists who report on it, because it reeks of high-tech innovation. It projects very clearly the idea — perhaps "fantasy" is a better word — that the earth's climate might someday be controlled by devices that are as cool and well engineered as a Porsche. It made the big, kludgy machine that I'd seen in David Keith's lab seem like a primitive contraption. But, of course, Keith's machine actually existed. Salter's was only a drawing.

Milo was obviously taken with it, too. "Cool!"

"Do you notice one thing that the boat is missing?" Salter asked.

Milo looked for a moment. "Sailors."

"Exactly! The boats are unmanned. They will operate entirely by remote control."

This excited Milo even more, since, like most boys, he loves remote-control devices.

"How many boats will there be?"

"Well, that depends. Right now, I'm thinking we might need a fleet of about three hundred to reverse the damage done to the climate since preindustrial times. We'll deploy them in different parts of the world, depending how much cooling is needed."

"How much will they cost to build?" Milo asked.

"About three million dollars each."

Milo nodded, as if that were entirely reasonable. "So how do they make clouds?"

"They don't *make* clouds," Salter replied carefully. "They *brighten* them."

Salter pointed to the Flettner rotors in the picture and said, "Inside each one of them is a sprayer, which will spray billions of tiny droplets of seawater into the sky. Some of those droplets—we don't know exactly how many—will be lifted up into the clouds, where the salt particles will act as cloud condensation nuclei. The water droplets that form on these particles will be smaller than the droplets that occur naturally in the clouds. And since clouds—at least these low-lying clouds we're talking about over the ocean—are really nothing more than water droplets in the sky, what happens if you make those droplets smaller?"

Milo thought for a moment. "They get brighter?"

"Yes!"

"So this is how you're going to save the planet?" Milo asked skeptically.

"Well, this is one idea," Salter said, hesitating only slightly. "We still have a lot of work to do."

Artists have always loved clouds, especially in the West. All over Europe, churches are frescoed with billowing clouds, symbolic of the mystery and power of God. In Renaissance paintings, clouds are often used as platforms for various deities (the Madonna and the baby Jesus are sometimes presented on a silvery bed of vapors), as well as to suggest a visible boundary between heaven and earth. In some Renaissance paintings, such as *Jupiter and Io* by Antonio Correggio, clouds are animate—Correggio portrays Io frolicking with Jupiter, who has taken the form of a very phallic-looking cumulus. Sixteenth- and seventeenth-century Dutch painters loved the sky, too, but they were more interested in texture and light than allegory and symbolism. In fact, you could argue that no one knew more about albedo engineering than Jan Vermeer, whose paintings, such as *View of Delft*, are precise studies in the way clouds shape and filter sunlight.

Western science, however, has not had such an easy time with clouds. It's not just that clouds are complex and difficult to measure; it's also that they are high in the sky, and until we developed a means of getting up into the clouds, they were pretty tough to investigate. The invention of hot-air ballooning in the late eighteenth and early nineteenth centuries helped some, awakening an interest in the mysteries of barometric pressure, wind and weather systems, and cloud physics. In 1803, British chemist Luke Howard, often considered the world's first meteorologist, published his now classic classification of atmospheric phenomena, *Essay on the Modifications of Clouds*, which was the first attempt to classify clouds into four basic types (cumulus, stratus, cirrus, and nimbus). German writer Johann Wolfgang von Goethe celebrated Howard as "the first to define conceptually the airy and ever-changing forms of clouds, thus delimiting and fixing what had always been ephemeral and intangible, by accurate observation and naming."

The study of clouds progressed modestly during the nineteenth century, driven mostly by an interest in lightning and the rise of the new science of meteorology. But as the popularity of the rainmakers in the late nineteenth century demonstrated, our ignorance of how clouds work remained vast. In part it was because the complexity of cloud physics was beyond the measurement of primitive scientific instruments; in part because, hot-air balloons aside, clouds were still too inaccessible to study.

That changed, of course, when the Wright brothers took flight at Kitty Hawk and the age of aviation began. Suddenly, people were spending a lot of time up in the clouds, and the development and advancement of flight depended on gaining a better understanding of the earth's atmosphere, as well as very specific weather-related phenomena, such as turbulence and ice formation in clouds. This was not just an issue for commercial aviation but also a matter of national security. During World War I, and even more during World War II, military control of the skies was often crucial to victory on the field below. And nothing bedeviled military strategists like clouds. "The most significant obstacle to bombing accuracy was cloud cover," one military historian wrote. The difficulty—and the necessity—of flying in all kinds of weather, and under all kinds of conditions, led to a number of important inventions, such as radio direction finders and radar.

One of the centers of wartime research into weather and aviation was the General Electric research laboratory in Schenectady, New York. GE, which had been founded by Thomas Edison, was in its prime during World War II. It was the industrial equivalent of Google today—the place where all the smart people wanted to work. One of the top scientists at the lab was Irving Langmuir, who had won the 1932 Nobel Prize in chemistry and was considered one of the great applied scientists of his day.

During World War II, Langmuir worked on a number of projects at GE, the most infamous of which was seeding clouds to make rain. With the help of an ambitious young assistant named Vincent Schaefer and a scientist named Bernie Vonnegut (brother of novelist Kurt Vonnegut, who also worked at GE, in press relations), Langmuir discovered that by adding solid carbon dioxide (dry ice) to clouds, he could encourage ice formation, which is one of the main processes by which water droplets develop into large enough particles to fall as precipitation. In late 1946, Langmuir and Schaefer undertook several high-profile cloud-seeding experiments in the

Berkshires of Massachusetts, dumping dry ice out of a small private plane into a bank of clouds below. At first glance, the experiments seemed to be a great success. In his laboratory notebook, Schaefer later wrote, "I shouted to [the pilot, Curt] to swing around and, as we did, we passed through a mass of glistening snow crystals . . . I turned to Curt and as we shook hands I said, 'We did it!' Needless to say we were quite excited. The rapidity with which the CO_2 dispensed from the window seemed to affect the clouds was amazing. It seemed as though [the cloud] almost exploded." Two days later, the *New York Times* ran a story about the experiment, explaining that "numerous practical applications" were expected to emerge from it, including "storage of moisture in the winter for spring irrigation and water power programs, steering heavy snowfalls away from city areas, and providing snow for winter resorts."

Before long, Langmuir and his colleagues discovered that other chemicals, such as silver iodide, also could initiate ice and provoke precipitation. He began to speculate openly about banishing drought, flood, and hail; controlling snowstorms; and turning the American Southwest into a garden of plenty. "Until the day he died, 10 years later," wrote one journalist, "Langmuir believed . . . he had witnessed a major turning point in the history of the world."

Of course, we know now that Langmuir was wrong. Cloud seeding has not changed the world. In fact, debate still rages about whether it actually increases precipitation at all. The main problem is the same one that rainmakers such as Charles Hatfield faced nearly one hundred years ago: how do we distinguish between natural rain—rain that would have fallen anyway—and rain that is artificially produced?

There is no doubt that in controlled laboratory experiments, chemicals such as silver iodide can cause ice formation and precipitation. And out in the real world, it can sometimes cause a local drizzle or snow to fall from simple stratiform clouds, such as what Schaefer and Langmuir witnessed. But these clouds are easy to manipulate, and rainfall or snowfall from them often evaporates before it hits the ground. The big, stratocumulus clouds that can

dump large amounts of water and snow are so complex and local-
ized that even experienced cloud seeders equipped with high-tech
instruments have a hard time hitting the sweet spot that will trigger rain or snow. On the right day, with the right cloud, with luck,
you might be able to seed a cloud and wring out a few extra drops of
rain. But far from changing the world, Schaefer and Langmuir really proved just how tricky it is to play with clouds and expect them
to react in predictable ways.

In the late 1980s, when most scientists were waking up to the dangers
of global warming, John Latham slept through the alarm. Latham,
who was in his late forties at the time, was a respected cloud physicist at the University of Manchester in England, best known for his
attempts to decipher the precise mechanism for how clouds generate lightning. It was not a simple question. After more than three
decades of work, Latham knew that it had to do with billions of hail
particles and ice crystals bumping into each other within the cloud,
but beyond that, much was still unknown. To Latham, the notion
that something as temporal and insubstantial as clouds could generate something as awesome and powerful as lightning was one of
the essential mysteries of nature. "For me, it's hard to imagine anything more amazing about clouds than the fact that little particles
of ice skating across each other can generate enough electric charge
to produce a spark ten miles long, with a hundred million volts
between its extremities," Latham told me. Compared to this, how
clouds might be involved in heating and cooling the planet was a
distant and abstract question, and one that, at the time, he paid little attention to.

Latham was raised in the little village of Frodsham, in northern
England. His father was an electrician. "No one in my family was
significantly educated or had been to university," Latham told me.
"I never dreamed of becoming a scientist. I thought I'd be a soc-
cer player or cricketer." In the aftermath of World War II—Latham
was eight years old when the war ended—the newly installed La-
bor Party government instituted programs that made it possible for

kids like Latham to get an education. His first love was English literature, but in his last years of high school, he took a science class and
got hooked. In 1958, he graduated from the University of London
with a degree in physics, then stayed on to get a Ph.D. in cloud physics. "It was an accident that I got into cloud physics," Latham told
me. "I was trying to dodge a special course in electronics, so took
meteorology instead, thinking it was to do with the stars." He went
to the University of Manchester after he received his Ph.D., eventually setting up the university's atmospheric physics department.

By the late 1980s, Latham had had enough of academic life and
decided to retire from his position at the University of Manchester. It was a time of personal turmoil for him—his marriage had
ended after he and his wife had raised four children, all of whom
were grown. He had also become impatient with academic bureau-
cracy. Although he wanted to continue his work on clouds, he had
other things he wanted to do with his life, including writing poetry,
which had always been a passion of his. After he retired, he thought,
he could make extra money with guest lectures and research grants,
as well as have more time to write and explore other interests.

But Latham soon discovered that earning money was not so easy
for an independent scientist. Adding to the pressure was the fact
that his kids—all of whom now had families of their own—were
struggling. "I was fairly desperate to supplement my income,"
Latham said. A fellow cloud scientist suggested that he contact the
newly formed Hadley Centre, the Met Office's climate research center in southern England. "Maybe you can get a consultancy there,"
his friend told him.

"If I was going to get the job, I knew I needed something new
to talk about," Latham said. As he was casting around for material,
he chanced on a paper by Anthony Slingo, a British climate modeler, about how changes in the concentration of nuclei—the tiny
particles that serve as gathering places for water droplets—could
brighten or darken clouds, especially over the ocean. Brighter
clouds reflected more sunlight and thus had a cooling effect on the

climate. Slingo argued that by increasing the number of low clouds (which are mostly over the oceans) by about 20 percent, you could offset the heat trapped by a doubling of CO_2 emissions.

The relationship between clouds and global warming was not something that Latham had ever given much thought to. "In fact," he told me, "Tony's paper was the only thing I'd read on the subject." But after reading Slingo's paper, Latham wondered whether deliberately changing the reflectivity of marine clouds could be part of the answer to global warming. Not that he thought much about global warming either. "It was something I was beginning to hear talked about," Latham recalled. Still, an obvious question arose: if the planet was overheating, why not just reflect away some of the heat?

This was more of a conceptual than a scientific breakthrough. Back in the 1960s and 1970s, when many cities were choked with smog and air pollution was a hot-button political and public health issue, scientists began to explore the impact of pollution not just on our lungs but also on the earth's atmosphere. Sean Twomey, an atmospheric scientist at the University of Arizona, wondered what impact pollution has on clouds and how clouds might contribute to the warming or cooling of the planet. It was well-known that sulfur dioxide, one of the main pollutants from burning fossil fuels, oxidizes in the atmosphere to create tiny sulfate particles and that these particles act as condensation nuclei for stray molecules of water, as well as traces of heavy metals, organic materials, and whatever else is around. Sometimes the droplets that form around the nuclei remain small and eventually evaporate; other times they bump into other droplets, growing larger and larger until they fall out as rain or snow. That much was known. What was not known was the impact these droplets have on the reflectivity of clouds. In 1977, Twomey showed that the amount of sunlight a cloud reflects depends not only on the number of droplets but also on their size. Clouds with lots of small droplets have more surface area than clouds with fewer big droplets (even if the amount of moisture is the same), and thus reflect more sunlight. Since pollution generally increases the number of small particles in clouds, wrote Spencer Weart in *The Discovery of Global Warming,* Twomey calculated that "the net effect of human pollution should be to cool the Earth."

In the 1980s, satellite photos of the earth reinforced Twomey's findings. You could see what looked like contrails over the shipping lanes in the oceans. They were in fact trails of clouds formed by the particles being spewed out by the diesel engines of ships. These ship tracks, scientists now know, have a pronounced cooling effect on the planet simply because the tops of the clouds are far lighter than the dark surface of the oceans.

After reading Slingo's paper, Latham thought, Why not manipulate clouds to cool off the planet? "I did some very rudimentary calculations in a few hours," Latham told me. "Any self-respecting modeler would commit suicide, it was so primitive." Basically, Latham concluded that a doubling of CO_2 emissions could be offset simply by doubling the concentration of condensation nuclei in low-level marine clouds. The point was not to manufacture clouds — it was to brighten them by adding smaller nuclei. Furthermore, Latham concluded that because the number of nuclei in the air over the ocean was relatively low compared to the number over land, fewer particles would have to be injected to double the concentration. He didn't speculate about how these particles might be created and lofted into the sky, but he did suggest finding a way to mimic the natural process of bubble bursting that happens in waves, in which tiny droplets of seawater are shot into the air. The seawater contains salt particles, which naturally act as condensation nuclei.

"The essence of the idea was all there before Latham came along," said Alan Gadian, a cloud physicist at the University of Leeds in England. "But no one had thought to put albedo manipulation of clouds into that context before — and to actually propose how it might be done. It was an idea that was way ahead of its time." In retrospect, Latham wasn't trying to revolutionize anyone's thinking about how to fix global warming. He was just looking for something to talk to the Hadley Centre about. "My motives were entirely mercenary," he recalled, not entirely joking.

Latham's idea was published in 1990 in *Nature*, a prestigious science journal, with the provocative title "Control of Global Warming?" Latham braced himself for a controversial response, but in fact the article landed with a thud. "I got two letters from the U.S., saying that what I was proposing was utterly immoral, and that if I really cared about global warming, I should be kicking in President George Bush's door and tell him to stop burning fossil fuels," Latham told me. And that was pretty much it. Thanks in part to his cloud-brightening idea, Latham did end up getting the consulting job at the Hadley Centre. But in the world of cloud physics, there were plenty of other interesting problems to think about. For the next decade, Latham hardly gave cloud brightening another thought.

In 2003, a colleague mentioned Latham's cloud-brightening idea to Stephen Salter, suggesting that the two men might want to get in touch. In the intervening years, Latham had taken a job at the National Center for Atmospheric Research in Boulder, Colorado, where he was continuing his work on lightning, among other things, and was living alone in a rough cabin in the Rocky Mountains with a big view of the sky. The two struck up an email and phone conversation and discovered they had much in common.

At the time, Salter was busy inventing a rainmaker. But this was not a nineteenth-century voodoo operation involving mysterious chemicals or massive bonfires. He had designed what looked like a giant mechanical eggbeater that could be installed a few miles offshore of drought-stricken areas. The eggbeater blades, blown by the winds, would spin around like a turbine. Except that instead of generating electricity, the eggbeater, which was more than a hundred feet tall, would lift seawater through a central tube, then shoot it out through the spinning blades in a fine mist. The idea was to increase the evaporation rate of seawater—essentially pumping water into the clouds, which would then be blown over the land, eventually dropping their loads. "The successful large-scale deployment of these rainmakers could reduce the number of people who are short of water by several billion," Salter told one reporter in 2002.

A number of atmospheric scientists told Salter that these machines would never work—the physics of rainmaking is far too complex to be affected by simply throwing water into the sky. One of the few scientists who was impressed with Salter's engineering prowess was Latham, who understood that a device that could lift seawater into the sky might be useful for his cloud-brightening idea. Was Salter interested in collaborating on the design of a cloud-brightening machine?

He was. But he realized immediately that a cloud-brightening machine presented some serious engineering difficulties. Among them was the fact that the machine would need to be mobile, allowing it to be moved as conditions changed to areas with the best opportunity for cloud manipulation. Even more problematic was the size of the droplets that Latham wanted to inject into the sky. To effectively brighten clouds, droplets needed to be extremely small—less than a micron in diameter. A human hair is roughly seventy microns in diameter. Bacteria are two microns.

So this was the challenge: Salter needed to design some kind of mobile device—a boat, most likely—that could pump seawater out of the ocean and then spray not thousands, not millions, but *billions* of tiny droplets into the sky. It also needed to do this in a highly efficient way, since the boat would need to generate its own power (refueling at sea would be complicated and expensive). The device also needed to be able to withstand rough seas and high winds and be as maintenance-free as possible, even when it was immersed for months in corrosive salt water. And the entire thing needed to be able to operate reliably without any human beings onboard, by remote control. Oh, and one other thing: the design had to be relatively cheap and easily replicable, because in order to have any real impact on the planet, you wouldn't need just one or two boats. You'd need hundreds.

At first, Salter was undaunted. He understood there were two separate design components: the boat itself and the sprayers. "The first thing I did was look at the energy flow," he told me. In other words, he first had to figure out how much energy he would need to move the boat around and power the sprayers, then he'd consider the options for generating that power. That would dictate all the other design parameters.

Figuring out how to generate the power to run the sprayers was simple enough. Salter decided early on that the boat could drag a turbine in the water—essentially a big fan under the hull that would spin a generator onboard, creating electricity. But how would he power the boat itself?

Given his background, Salter first looked at drawing power from waves. He quickly discarded that idea—wave power, he concluded, makes sense only for stationary devices. Conventional sails, with their riggings and masts, were too complex. He looked at wing sails, which operate essentially like vertical airplane wings, but realized they might get in the way of the sprayers. Finally, he came up with the idea of using Flettner rotors, named after the German engineer Anton Flettner, who first installed vertical rotors on a sailing ship in the 1920s.

The rotors may look like giant toilet paper tubes, but they are actually more efficient than conventional sails. They work by exploiting a phenomenon called the Magnus effect, in which a spinning object creates a whirlpool around itself and a force perpendicular to the line of motion (in this case, the wind). It's the same phenomenon that causes a spinning golf ball to slice to one side, and harnessed on a large scale, it can be an effective way to power a boat. In 1926, a boat equipped with Flettner rotors sailed across the Atlantic without any problems. With the coming of diesel engines, however, the technology never caught on and was largely forgotten by engineers until Salter resurrected it.

All in all, it took Salter only a couple of months to sketch out a basic design for the boat. It was simple, elegant, efficient, and, in its own way, revolutionary. Designing the sprayers, however, was another matter.

"Getting the particles the right size is by far the most important element in this whole scheme," Alan Gadian, who has looked closely at Latham and Salter's idea, told me. If the droplets are too big, they will glom on to one another. Instead of brightening the clouds, they will cause the clouds to rain out—essentially making them disappear and allowing more sunlight to hit the water, precisely the opposite effect of what the scientists intended. If the droplets are too small, they will simply evaporate and have no effect at all. According to Gadian, the margin for error is not large. Latham and others have determined that 0.8 micron would be the ideal size. Two microns is probably too large and would cause drizzle. One-half micron is likely to be too small. And not only does the size of the droplets have to be right, but for the sprayers to have any impact, billions of these particles would have to be created for hours—and likely days—at a time. It is a nanoscale engineering problem of colossal dimensions. In fact, nobody has ever made liquid particles this size for any purpose—much less for manipulating clouds on the high seas.

Salter's first idea was to use pond foggers, which use high-frequency vibrations to release water molecules in the form of mist. His second idea involved bouncing droplets back and forth between electrically charged plates. Then there was a scheme involving the collision of high-pressure jets precharged with compressed air, which would cause drops of water to collide and spatter. None of these was satisfactory—they were all too unreliable or too difficult to control, or they required too much energy.

The simplest mechanism involved micro-nozzles. Micro-nozzles are really nothing more than a bunch of holes in a flat plate—water is forced through, just as in a garden hose—except in this case, the holes are impossibly small. And, compared to the alternatives, they are very energy-efficient. Salter initially rejected this for an obvious and nontrivial reason: there is a lot of junk in seawater, from plastic bags to microscopic organisms. Even with a good filtering system, the nozzles would likely get clogged over time. But then at a geoengineering conference in California in 2006, Lowell Wood suggested that Salter reconsider the idea. Wood pointed out that filters to desalinize seawater are fine enough to take out all the silt and microorganisms—they can even screen out the polio virus, Wood argued, which is about thirty times smaller than the droplets Salter wanted to make. And they are reliable: "People die if they fail," Salter said.

So Salter went back to the idea of micro-nozzles, using an ultra-filtration system similar to the one Wood recommended and designing a way to back-flush the nozzles—that is, rinse them out with clean water—on a regular basis. He began a collaboration with Tom Stevenson, the operations manager at the Scottish Microelectronics Centre at the University of Edinburgh, to design and fabricate micro-nozzles out of silicon wafers.

Reading about this on the page, it all sounds straightforward and boringly technical. In fact, it wasn't until I visited Salter in Edinburgh and he handed me a silicon wafer that I grasped the fundamental complexity of what he was talking about. The wafer was nothing special—about eight inches around (about the size of a big chocolate chip cookie), a fraction of an inch thick, a shiny metallic surface on one side, a dull gray on the other. It was exactly the kind of wafer that is used to manufacture the computer chip that powers your laptop. Salter explained that the micro-nozzles would be etched into a wafer like this one, and that there would be six wafers in the center of each of the Flettner rotors, for a total of eighteen wafers per ship. The droplets would flow out of the hollow center of the rotors like smoke out of a chimney.

Holding the wafer in my hand, I asked idly, "How many micro-nozzles will be in each wafer?"

"About one point five billion," Salter said coolly. Then he added, as if to clarify, "Remember, they will be very small holes."

"Amazing," I said, somewhat lamely.

"Micro-fabrication people are not scared by big numbers."

"So with eighteen wafers on each ship . . ."

Salter was way ahead of me. "Something like thirty billion nozzles per ship," he said. "We need a lot, in case some get clogged up. And, you know, we need to put a lot of droplets up into the sky."

In Silicon Valley, I'd seen similar wafers many times, had toured the clean rooms where they are made, and had drunk beer with men who designed them. I knew about the complexity of growing and doping the silicon, the nanoscale etching of circuitry into the wafers, and, most important, how these wafers are cut up into the tiny silicon chips that control everything from our computers to the MRI machines that look into our bodies to the missiles that threaten and protect our cities. As I flipped the wafer over in my hand, so shiny and clean and beautifully engineered, I couldn't help thinking about how logical and how inevitable it was that the best hope of manipulating clouds, and perhaps the earth's climate, would come down to this same fundamental technology. It has become a cliché to say that Silicon Valley has changed the world. But if Latham and Salter's idea for brightening clouds ever works, it could literally be true.

There are three regions where low-lying stratocumulus clouds are more or less permanent fixtures over the ocean: off the coasts of California and Peru in the Pacific, and off the west coast of Africa in the Atlantic. One recent study suggested that if the number of cloud droplets in those regions alone were doubled, the clouds would be brightened enough to offset nearly half the warming that would result from a doubling of CO_2 concentrations in the atmosphere. To completely compensate for a doubling of CO_2, Latham believes, about 50 to 70 percent of the clouds that cover the world's oceans would have to be brightened. But we wouldn't have to take it that far. Unlike particles in the stratosphere, which would have a general cooling effect on the entire planet, cloud brightening could be deployed in particular regions, giving future geoengineers a more precise tool for customizing the climate. "You could use [it] to air-condition Abu Dhabi," Ken Caldeira said half-seriously. In theory, it could also be used to shift ocean circulation patterns, bring rain to drought-stricken regions, or even deflect or weaken hurricanes. (Hurricanes get their energy from the heat differential between the warm sea surface and the cool air above, so chilling the ocean even slightly could reduce a storm's ferocity.) Cloud brightening also has the virtue of being easily reversible: if weird weather patterns developed, we could shut off the machines, and everything would revert to the previous state almost instantly.

Of course, a fleet of satellite-controlled boats racing around the world in pursuit of clouds—commanded, presumably, from some bunker at Climate Control Central—and spewing billions of nanoscale droplets of seawater into the sky is difficult to imagine. Even aside from the obvious sci-fi aspects of this plan, there are certainly plenty of reasons to be skeptical.

The droplets, for example. Salter's micro-nozzles are still unproven, and although other researchers are working on different methods to create droplets, nobody has done it yet in the mass quantities that are required—much less in a marine environment. Even if we could solve that problem, how would we loft them into the clouds? Salter believes that we could rely on natural turbulence, but some cloud physicists are skeptical. One of the key processes in creating clouds in the marine boundary layer—the scientific term for the humid, turbulent, two-kilometer-deep part of the atmosphere over the oceans—is convection. At night, convection pulls up moisture from the warm ocean to form clouds at the top of the boundary layer. But as the top of the clouds heats up from the sun and the temperature difference disappears, that convection can weaken, making it much harder for the nuclei to rise into the clouds.

Of course, if the conditions aren't right, it doesn't matter whether the nuclei levitate or not. If the sky is too clear, there will be no clouds to brighten. If the sky is too dark and cloudy, the impact of the nuclei will be imperceptible. If the clouds are layered—clouds on top of clouds—brightening the clouds at lower altitudes won't have any impact on how much sunlight they reflect.

And if we *do* manage to get enough nuclei up into the clouds, then what? Marine boundary layer clouds seem, at first glance, pretty simple—especially compared to big convective clouds like the ones you see boiling on the horizon when a storm is approaching. Those clouds are often ten thousand feet deep, made of both water and ice, and are chaotic swirls of wind and currents. In contrast, marine boundary layer clouds are thin, relatively uniform, and made up entirely of water (no ice). "But these marine clouds are also more finely tuned, more fickle," Latham told me. For example, adding more nuclei might indeed brighten them, but it also might increase the evaporation rate within the clouds, thinning and disrupting them and perhaps causing them to disappear.

Even if we could engineer the cloud-brightening machines to work, there is a whole set of meta-questions about their impact on weather and climate. For instance, it's likely that the areas immediately downwind of a cloud-brightening operation would see less rainfall (smaller droplets are less likely to rain out). On a larger scale, temperature differences between oceans and land drive a number of different climate phenomena, from wind to monsoons. One of the advantages of injecting particles into the stratosphere to reflect sunlight is that the cooling effect would be more or less global. Cloud brightening, however, would cool a particular part of the ocean, increasing the land-sea temperature differential and perhaps altering the dynamics of the climate system. The cooler water could also change circulation patterns in the ocean, which could affect wind and rainfall far from the area where the cloud-brightening operation is taking place. For example, it is possible that brightening clouds off the coast of California could disturb the periodic rhythms of El Niño, an atmospheric phenomena that has a big impact on the amount of rainfall in the Pacific region.

One of the most vulnerable areas may be the Amazon rain forest. Andy Jones, a climate modeler at the Met Office Hadley Centre in England, has shown that cloud brightening could sharply decrease rainfall in the Amazon, which would be devastating to the region for a number of reasons, not least because the Amazon is one of the world's most important carbon storage sites. In contrast, Phil Rasch, a respected climate modeler at the Department of Energy's Pacific Northwest National Laboratory, has done similar modeling of the Latham-Salter proposal and found no evidence of serious effects in the Amazon. In a broader experiment, Ken Caldeira and his colleagues have found that if all ocean clouds were brightened, rainfall would likely decrease over the oceans but *increase* over land essentially everywhere, including the Amazon (due to changes in the flow of moist air over land). This disagreement in model results is not surprising, given how little work has been done to date and how difficult precipitation is to forecast accurately (in large part because of the complexity of clouds). More robust modeling studies might help settle this issue, but the best way to gain confidence in likely outcomes is to brighten a few clouds and see what happens.

On one level, you might think that spraying seawater droplets into some clouds would be a simple matter. After all, if we just want to see if we can create the right size droplets and loft them into the clouds, we don't need to launch a large-scale experiment that puts the Amazon at risk. We certainly don't need to build the fancy boats that Salter has designed. If he (or someone else) can come up with a way to spray tiny droplets of seawater into the air, we could install sprayers on the decks of a few ships, let them spray for a few weeks in a region that is ripe for manipulation, and see what happens. A carefully conducted experiment like that would be a great boon for scientists who study cloud dynamics. "We know we can perturb clouds," Rasch said. "The question is, can we do it where we want to and how we want to?" Maybe the clouds wouldn't brighten at all. Maybe they would brighten far more than expected. Whatever the results, the information that could be gleaned from such a deliberate experiment could go a long way toward determining whether cloud brightening is an idea worth further investigation.

Even if the idea failed completely, research that contributed to a greater understanding of how clouds work would help us assess the dangers we face from rising CO_2 levels. "Basically, our current understanding says that sensitivity to doubling [of CO_2 levels from the preindustrial state] might be as little as 2°C, or could be even as high as 10°C," climate researcher Ray Pierrehumbert wrote to me. "How aggressively we act to reduce CO_2, and the nature of the harms, depends on what the answer is. The main uncertainty is clouds, and it is only through monitoring of clouds that we will have a chance of tracking what the real climate sensitivity is, and get early warning of whether we're on a 'high sensitivity' track. The early warning idea might not actually be feasible, but without tracking clouds, it is clearly impossible."

In any case, actually conducting such a field test is a much more difficult matter than it would appear to be at first glance. From a purely scientific point of view, the whole endeavor is exceedingly complex, not just because we'd have to spray just the right number of nuclei into just the right kind of clouds at just the right moment. We'd also need to have the monitoring and observational equipment in place to detect any changes and to distinguish them from natural effects. As you might imagine, measuring the amount of sunlight reflecting off a cloud is not a trivial operation. Typically, a field experiment of this sort, which might take place over a one-hundred-square-kilometer area, would involve a number of airplanes and high-tech measuring devices, as well as scientists from universities and government organizations from several countries. And the whole enterprise would be neither quick nor cheap. Kelly Wanser, a Silicon Valley entrepreneur who is trying to raise funding for just such an experiment, believes that if everything went smoothly, it would take four years and about $20 million to run a field test for Latham and Salter's idea—and then several more years to interpret the data.

But the real roadblocks are likely to be political, not technological. The moment that such an experiment was announced, the ruckus would begin. *This is not about science,* opponents would say. *This is about hubris. This is about evil geoengineers messing with Mother Nature. This is the first step down the slippery slope toward full-scale deployment. This is a dangerous quick fix. This is technology run amuck.* It wouldn't matter that the experiment would be, by any objective standards, as natural as an organically grown carrot. After all, it would consist of nothing more than tossing tiny droplets of seawater into the air, mimicking an act that Mother Nature herself has been performing for a very long time. And it wouldn't matter that the experiment would be instantly reversible: in the unlikely event that brightening a few clouds did have some unforeseen impact, we could shut the sprayers off immediately, and the droplets would vanish. In a sense, the facts wouldn't matter, because to those people who are fundamentally opposed to geoengineering, deliberately messing with the climate is morally and ethically wrong, no matter what the scale, no matter what the impact. You can see how this could easily devolve into the climate equivalent of the abortion battle, fraught with symbolism, emotion, and political expediency.

There were already hints of this in the fight over iron fertilization (see chapter 7), as well as in the heated rhetoric that accompanied a report on geoengineering released by the British Royal Society in 2009. The study, undertaken by one of the oldest and most respected scientific organizations in the world, was a comprehensive attempt to consider the risks and benefits of a number of geoengineering options and to lay out a road map for further research. It caused a fury in some quarters. A few days before the report was made public, the ETC Group, which is based in Canada and describes itself as "dedicated to the conservation and sustainable advancement of cultural and ecological diversity and human rights," blasted the report, issuing a seven-page paper about the evils of geoengineering titled "The Emperor's New Climate: Geoengineering as 21st Century Fairytale." Many of the dangers that the ETC Group cited are real and substantial, but the paper was packaged in such a way as to turn it into an ad hominem attack on the scientists who participated in the Royal Society study. It suggested not only that all forms of geoengineering are bad but also that one should be tarred and feathered for talking openly about them: "The Royal Society will play an important role in this performance by offering a prestigious platform and global microphone to some modern-day tricksters."

The paper sparked a heated reply from some scientists, including Ken Caldeira: "It is one thing to suggest that we are uninformed, misinformed, or even deluded, but another thing entirely to suggest that we are acting with the intent to 'trick' people into doing things that might harm the environment. Charges that we are acting in bad faith are unfounded, reckless, repugnant, and malicious." In fact, if the ETC Group was really concerned about the dangers of geoengineering, Caldeira suggested, they should be advocating more research, not less. "If ETC really believes the emperor really has no clothes," Caldeira wrote, "they should be happy to join us in calls for an investigation into the emperor's clothing." These words are particularly cutting from Caldeira, who is about as far from the stereotypical image of a hubristic geoengineer as they come.

The challenge of field-testing doesn't just apply to cloud brightening, of course. Injecting particles into the stratosphere, even on the most modest scale, would be even more likely to incite a riot. It may be that in the coming years, as our failure to deal with global warming by traditional means becomes more apparent, the idea of messing with a few clouds will become less frightening. But maybe not. "All it takes is a few loud voices, and they can stall this for years," said David Victor, who has written frequently about the cultural and legal complexities of global warming. "Cloud brightening may be a good idea or a bad idea. But it may be that we will never know, because we may never have the chance to try it out."

The paradox, of course, is that if further research is thwarted, it doesn't necessarily mean that nobody will try to brighten clouds or shoot particles into the sky. It only means that the geoengineers who do end up trying it will be more ignorant of the real risks. "We know that putting a full right hand lock on the steering wheel gets us into the ditch on the right side of the road, while [a] full left hand lock gets us into a ditch on the left," Salter wrote in an email about the importance of pushing ahead with research, despite the obvious risks. "This is an argument for learning how to steer, not a reason for removing all steering wheels."

NINE

·····················

A Global Thermostat

WHILE REPORTING THIS BOOK, I was struck by how quickly geo-engineering moved from fringe science to the mainstream. In the fall of 2006, when I first visited David Keith, it was nearly impossible to discuss the subject without a giggle or a groan. A year later, it was the subject of a two-day conference at the historic American Academy of Arts and Sciences on the campus of Harvard University in Cambridge, Massachusetts. Many big names in the academic world were there, including Larry Summers, then an economics professor at Harvard and now the chief economic adviser to President Obama, as well as a number of scientists I knew very well, such as David Keith, Lowell Wood, and Ken Caldeira. The two-day conference, which was hosted by Keith and Daniel Schrag, a geochemist at Harvard, was described to me by one participant as a "coming out" meeting for geoengineering—in essence, a gathering of top scientists and thinkers to decide whether geoengineering was worth further research.

For Schrag, just hosting such a meeting at Harvard was risky business. Schrag is an ambitious academic who is as skilled at political maneuvering as he is at atmospheric chemistry. On the one hand, it never hurts to be seen as a leader in a hot new field like geoengineering. On the other hand, geoengineering was still the kind of subject that high-profile scientists and thinkers were not sure they wanted to be associated with (and having Lowell Wood in the room — who, brilliant as he may be, is a living symbol of Big Science Gone Wild — didn't help). Even worse, there was a whiff of imperialism about the whole affair: if one had a conspiratorial turn of mind, the meeting on the Harvard campus could easily be portrayed as a secret gathering of a new climate cabal. After all, the forty or so people attending the meeting — scientists, mostly, as well as a few economists — were all middle-aged, affluent elites. And all but one, Sallie Chisholm, an ocean researcher from MIT, were men. ("They invited me at the last minute, and I think it was because they suddenly realized the whole meeting was made up of white men," Chisholm told me later.)

The academy is located in a modernist building on the outskirts of the Harvard campus. When I arrived on the second day of the conference, I bumped into James Fleming, a science historian from Colby College in Maine who has written extensively about the history of weather and climate modification and is deeply skeptical about geoengineering's prospects. Fleming seemed dismayed about the gung-ho nature of the discussion at the conference. He was particularly taken aback, he told me, by Summers, who, according to Fleming, boasted about his role in persuading President Clinton not to sign the Kyoto Protocol, then went on to compare the need for geoengineering research with the need for fallout shelters during the Cold War. Fleming rolled his eyes and said, "He had no idea what he was talking about." (When I tried to confirm this with Summers shortly after the meeting, we traded a few emails, but as soon as I asked him specific questions about his views on geoengineering, he ended our correspondence.)

After I talked to Fleming, I headed into the bathroom, where I met a tall man with large, round glasses and white hair at the sink.

I didn't know him, but it was clear that he was attending the geo-engineering conference, and we introduced ourselves. His name is David Pritchard, a highly regarded physicist at MIT who has mentored several Nobel Prize winners. Pritchard seemed excited about the meeting, specifically citing a presentation by David Keith. (Pritchard had been Keith's thesis adviser at MIT.) Keith had talked about the "immense leverage" that geoengineering technologies give scientists to change the earth's climate, Pritchard explained. "This is a big deal," he said. "It has the potential to change everything. I don't think we've even begun to grasp the implications of this."

I agreed—although at that point in my research, I was still wondering whether the whole idea was just a bad sci-fi novel writ large.

As Pritchard dried his hands with a paper towel, he said, "I think you have to look at this in a larger historical perspective. The story of the twentieth century is about technology putting power in the hands of people. You see it with the computer revolution, of course. But you can also see it in other ways. A single person can now engineer a microbe that could kill millions of people. Terrorists can use a jet to crash into the World Trade Center. Now one nation, or even one person, can manipulate the entire earth's climate."

Pritchard tossed the crumpled towel into the wastebasket. I told him that I'd never seen the personal computer, terrorism, and geo-engineering as part of the same evolutionary trend, but it obviously made sense.

"Geoengineering is just a tool, and tools can be used for good or evil," he said. "How do you stop someone—or some state—from trying to take control of the climate? That is a very interesting question. And then you have countries like Russia, which actually like global warming, because they want to get at the oil and gas in the Arctic. How would they react to someone trying to cool the planet?"

"You can certainly see the potential for conflict," I said.

"Yes, you can," Pritchard agreed. "I remember going to conferences years ago where we talked a lot about whether moral progress keeps up with technological progress. This reminds me a lot of that."

Then he wished me well and headed out the door.

Since I talked with Pritchard that day, I've had hundreds of conversations with smart people about the complexities and consequences of geoengineering. And yet that encounter sticks in my mind. Why? In part it is because Pritchard summed up the central governance question about geoengineering: If we begin to engineer the climate, whose hand will be on the thermostat? And how do you stop a lone actor—armed with good intentions or bad—from screwing up the climate for all of us?

But Pritchard also helped me understand a larger point: the politics of geoengineering are just as complex as the technology. The simple truth is, we have crossed over an important dividing line that separates us from all the billions of people who came before us. We may not be morally more sophisticated or smarter or better artists or bolder scientists or more loving parents, but we do have one thing that no civilization before us has ever had: we have the power to intentionally change the climate of the planet we live on. We might do it badly, we might do it well; we might do it quickly, we might do it slowly. But we can do it. What matters now is the human part of the equation.

Are we up to the challenge?

It's no accident that a guy like Pritchard was able to put his finger on the central challenge posed by geoengineering. As a young scientist in the 1960s, Pritchard studied under the last of what David Keith calls the "bomb johnnies" in the world of physics—that is, scientists who came of age in the nuclear era, when all the brightest minds went into bomb building. And there are a lot of similarities between nuclear bombs and geoengineering, including the fact that both are big, technocratic operations that can potentially put an end to civilization as we know it.

From a governance point of view, they also have a lot in common. Just as the central question of the nuclear age was how to keep a Dr. Strangelove from pushing the button, the central question of the geoengineering age will be how to prevent a Dr. Strangelove from hacking the climate. And although geoengineering schemes don't have the shock and awe of a nuclear attack, the stakes are, if possible, even higher. A nuclear bomb dropped on a major city might kill millions of people. A geoengineering scheme gone awry could make the planet uninhabitable for billions. The barriers to entry are also much lower for geoengineering: you don't need any MIT-educated physicists or rare materials such as plutonium to muck around with the climate. Compared to a nuclear weapons program, throwing dust into the stratosphere is easy.

In the world of climate policy, geoengineering requires a lot of new thinking. Our approach to dealing with global warming so far has essentially been to ask everyone on the planet to come together, understand what is at stake, and do the right thing. Let's put aside greed and petty ambition, we say, and rally for the cause of humanity and the Amazon rain forest and the polar bears, monarch butterflies, and millions of other creatures that we have never met and whose names we can't pronounce but that will all go extinct if we don't give up our SUVs, move to smaller houses, and quit flying off on vacations to the Caribbean. This "come together and save the world" strategy is embodied in the Kyoto Protocol, which was initially adopted back in 1997, went into effect in 2005, and currently has 183 nations signed on—virtually everyone on the planet except the country that matters most, the United States (even China signed the agreement, although as a developing nation, it has no legal obligation to cut emissions). The net effect of all this, as far as the atmosphere is concerned, has been zero. Not only have we not reduced global CO_2 emissions, but they are continuing to rise faster than ever.

What's happening to our climate is known, in political terms, as a tragedy of the commons. Even when most of the farmers in a village agree not to overgraze their community pasture, one or two let their sheep out in the middle of the night, thinking they will gain an advantage over the others and make a little extra money. Before long, the pasture is ruined, the sheep starve, and all the farmers have to find new work as telemarketers.

But Kyoto-like agreements that address the problem of dumping CO_2 into the atmosphere have little impact on the issues raised by geoengineering. In fact, international law and policy experts have — not surprisingly — given very little thought to geoengineering, and that is a big problem, according to David Victor, who has written several influential articles about the complexities of geoengineering governance. So I asked him, what is his nightmare scenario? "Unilateral deployment, at a large scale, with untested technology," he replied.

What might such a scenario look like?

Scenario 1. It's the year 2030. The Himalayan glaciers that provide fresh drinking water to 40 percent of the world's population are vanishing. In China, crops are failing because of drought, leading to food shortages for millions of people. Citizens are taking to the streets. The Chinese government is under political pressure to do something. Building new dams and aqueducts is possible, but it will take decades and might not work. Shutting down coal plants — even if it were feasible, which, politically and economically speaking, it is not — would have little immediate impact. Instead, Chinese officials decide to inject particles into the stratosphere over the Himalayas to deflect sunlight and cool the glaciers. They might do it with artillery, high-altitude balloons, or military jets. The operation could be started in secret (other nations would detect it fairly soon), although for political purposes, it would likely be staged to show the nation's leaders going to heroic efforts to fix the problem.

The scheme may or may not preserve the Himalayan glaciers. The larger trouble is, particles shot into the stratosphere over the Himalayas don't stay over the Himalayas — so without exactly intending to, the Chinese end up starting a global geoengineering program. What is the rest of the world going to do about it? What if the Asian monsoon shifts northward or suddenly delivers significantly less rainfall. Does India blame China? Does the lack of rainfall exacerbate tensions between India and Pakistan? And how will

Russia, whose leaders are eager to exploit the melting Arctic, feel about the idea of the Chinese deciding to cool the planet? However, suddenly stopping a particle injection program has its own risks (due to the rebound effect discussed in chapter 6). "In some ways, it could be more dangerous to stop a kludgy geoengineering operation than to continue it," Victor told me. In other words, if China does it badly, we may decide that it's better to keep it going and try to fix it than to strong-arm China to shut it down (as if that were really an option).

Scenario 2. Developing nations are tired of the dithering and double talk from big polluters about cutting emissions. Climate impacts are hitting them the hardest: island waters are rising; drought is parching Africa; anger at Western nations and China rages. With the help of rich, sympathetic backers, developing nations band together and threaten to cool off the planet themselves if the West doesn't cut CO_2 pollution by 10 percent each year. The West says it's impossible and calls their bluff. To save face, the developing nations begin injecting particles into the stratosphere using airplanes and rockets and artillery shells and slingshots and whatever else they have on hand. Complications ensue.

Scenario 3. According to *Forbes,* there are about one thousand billionaires in the world. It's not hard to imagine that one of them—or a group of them, for that matter—might decide to take it upon himself or herself to save the planet through geoengineering. Victor coined a name for these types: "Greenfingers," a play on the James Bond character Goldfinger. "Everyone immediately thinks of the Google guys or Richard Branson, but in fact the world is full of billionaires," Victor told me. "Some of them are living in places where they might not feel the same taboos about messing with the atmosphere that people living in highly transparent, democratized countries might." And they don't have to actually own a fleet of high-altitude aircraft to make it happen. They could use rockets based in friendly territory—indeed, some places might welcome the investment, just as they have welcomed tax haven status. Or, in theory, they could fund the science, buy off the politicians, and launch a PR campaign promoting the idea—all with the best intentions, convinced they are taking a heroic step into the future. Imagine T. Boone Pickens pushing geoengineering instead of natural gas. Imagine Steve Jobs building a cloud-brightening machine.

Nobody really knows how realistic any of these scenarios is simply because no one has tried to do any of this yet. One thing to note about them, however: they all involve solar radiation management technologies—mostly stratospheric aerosols. Right now, that is the only technology that might be deployed fast enough, and cheaply enough, to pose an immediate threat. And even then, you can't just load a few bags of particles into a 747, fly up into the stratosphere, and chill the planet overnight. Even if you resolved the basic engineering issues about how to create and disperse the particles, it would take months to actually have any impact on the climate. So nobody worries about a Kim Jong Il launching a geoengineering project tomorrow. The threat is longer-term and more insidious. Unlike nuclear or biological weapons, geoengineering is not about annihilation. It is about dominance and control.

Whenever a new technology emerges that challenges the existing moral and political order, public debate swings from one hysterical pole to the other. Robotics will revolutionize industry or destroy the human soul. Genetically modified crops will feed the world or lead to Frankenfood. Nanotechnology will be the solution to our energy problems, allowing us to build highly efficient solar panels and transmission lines, or end up dissolving the world into gray goo. Adding to the complexity of all this is the fact that there are probably ten people in the world who are wise enough and knowledgeable enough to make a judgment about, say, the risks of nanotechnology run amuck. I have no idea who they are, and I bet you don't either.

In the midst of this swirl of fact and fantasy that surrounds new technology, the easiest—and, in some ways, the sanest—response is simply to say, *Stop.* Do we really need to clone cats? Can't we feed

the world without inserting fish genes into tomatoes? Is augment-
ing our intelligence by implanting a microchip in our brains really
the surest route to wisdom? "In general, I think we should count
ourselves the lucky inheritors of 500 years of scientific progress,"
writer and climate activist Bill McKibben has said. "We live long,
comfortable, easy lives in the West. The question is, Should we keep
pressing endlessly forward or decide to rule out certain quantum
leaps?"

This is an excellent question. So for the sake of argument, let's say
world leaders come together and decide that the risks of many geoengineering schemes are just too great and we should ban any attempts to deliberately modify the climate in order to offset global
warming. How would we institute such a ban?

The most obvious way to prohibit unwanted behavior in a civilized society is to pass laws against it. That's easy enough to accom-
plish if the goal is to prevent someone from stealing someone else's
car. But when it comes to geoengineering, there are a number of
problems, starting with the fact that, in legal terms, geoengineering is not easy to define. Is it okay to throw seawater into the air to
brighten one cloud but not one hundred? What is the difference
between injecting particles into the stratosphere with a hose connected to a high-altitude balloon and injecting particles into the
lower atmosphere by burning coal? When you start drilling down,
you hit questions about scale and, more important, intention. We
have come to grudgingly accept particles injected into the atmosphere as a byproduct of creating electricity, but injecting particles
into the stratosphere to cool the planet strikes many of us as hubristic and dangerous.

Another problem is that we don't know enough about the real
risks of geoengineering to formulate an intelligent global agreement. If we attempt to limit experiments based on the risks they
entail, how do we make a judgment about which risks are worth
taking? To write laws and regulations governing an emerging technology, we need to know something concrete about it. "Legislation
based on vague generalizations is vague—and that makes it use-

less," Victor told me. Case in point: nuclear weapons. It took decades before the dangers and political complexities were understood well enough to craft strong agreements to stop nuclear proliferation.

There is only one international treaty that is even remotely related to geoengineering, and that is the Environmental Modification Convention, signed in 1977. That agreement, which prohibits the hostile use of "environmental modification techniques" such as cloud seeding, was inspired by the fact that militaries were considering weather modification as a weapon of war. But cloud seeding never worked as a weapon of war, and the treaty, widely viewed as weak and ineffective, has never really been tested. Another treaty focused on restraining people from messing with the global commons, the London Convention, has worked well to stop the seabed disposal of radioactive waste. But it didn't do much to slow down Planktos's plans to dump iron into the ocean (see chapter 7), and, at the behest of corporate interests and environmental groups, it has been revised to allow for the disposal of CO_2 deep under the sea, in the seabed.

Then there is the question of enforcement. The UN Security Council has no sovereignty over individual actors in individual nations, and many of the major nations that would be most likely to launch a geoengineering project have veto power on the council. Other possible enforcement actions, such as trade sanctions and tariffs, aren't going to stop a nation hell-bent on geoengineering. Here in the United States, we can pass laws saying "Thou shalt not fill the stratosphere with particles," but they're not going to carry a lot of weight in Russia or Brazil. Ideally, the governments of Russia and Brazil will pass their own laws, and a global consensus will emerge. But what if it doesn't?

During one of our discussions about geoengineering governance, Keith sent me an email that underscores the complexity of the problems we face:

Suppose India begins a decade-long well-funded Solar Radiation Management (SRM) research program that moves systematically from modeling to micro-scale deployment tests. Suppose that program operates with complete transparency and is governed by an advisory body that includes the world's leading scientists along with retired politicians like Nelson Mandela, Bill Clinton, and Mikhail Gorbachev. Suppose that India then begins a gradual well-monitored program of SRM deployment arguing that it has exhausted all other avenues of action to limit climate impacts. Such action would effectively seize the initiative on global climate control and would be very hard to restrain even if there was an early international agreement banning geoengineering.

Trying to ban geoengineering right now is not a good idea — not just because it is likely to be ineffective but also because we simply don't know enough about the real risks to set up a coherent and binding agreement. "It would be so vague as to be irrelevant," David Victor said. "Or, worse, it would discourage the most responsible geoengineers from developing technologies and testing the ideas in ways that are responsive to international norms." Instead, responsible scientists should be encouraged to experiment with various technologies so that their true risks can be discovered. It goes without saying that any geoengineering experiments need to be carried out with the highest degree of accountability and transparency, as well as an acknowledgment that there is a slippery slope from experimentation to deployment.

Still, the greatest dangers, as always, are from ignorance. "Troublesome and high-risk technology should be treated like rogue states," wrote Kevin Kelly, one of the founding editors of *Wired* and author of *Out of Control: The New Biology of Machines, Social Systems, and the Economic World.* "What you don't want to do is banish and isolate them. You want to work with the bully and problem child . . . Prohibiting them only drives them to the underground, where their worst traits are emphasized."

David Victor believes that openness and collaboration are vital. In his view, an intensifying sequence of conferences, research projects, data sharing, and brainstorming will eventually evolve into the development of scientific norms and best practices. In the 1970s, biologists gathered at the Asilomar Conference Grounds in California to discuss self-regulation of the emerging field of biotechnology. The Asilomar conference was a great example of scientists in a controversial area coming together to put limits on their research. A similar effort is clearly called for with geoengineering, and as of this writing, several such conferences are in the works. In the coming years, some scientists expect to see the emergence of an ad hoc geoengineering governing body similar to the Internet Engineering Task Force, which oversees the technical development of the Internet through an ever-evolving collaboration of scientists and engineers. A group like this devoted to geoengineering, if it were open and international in scope, could lead to more coherent and informed international agreements.

"The point of all this is not to stop geoengineering," Victor told me. "That is never going to happen. The point is to shift the odds, so when someone does it, they do it in a responsible way."

A simple way to understand the governance challenges in a geoengineered world is to imagine a big old house with several different living spaces in various states of repair. On the ground floors, in the most spacious and well-kept rooms, the rich people live. On the upper floors, where the rooms are smaller and less luxurious but far more numerous, the middle-class people live. Stuffed down in the basement, up in the attic, and in a leaky, unheated ramshackle wing out back are the masses of poor people.

Now imagine that this old house has a single thermostat and a heating system only in the main part of the house. If the house is in, say, my hometown in Silicon Valley, which has a Mediterranean climate, there probably won't be much conflict over where to set the dial. You can imagine that in December and January, when the temperature falls to 30 degrees Fahrenheit or so, the people down in the basement might want a little more heat than the rich people in the well-insulated rooms, and the people in the ramshackle wing might be unhappy and shivering. But for the most part, the potential for conflict will be modest.

But think of this same house in northern Canada. On the one hand, if you set the thermostat to a comfortable temperature for the privileged few, the people in the basement and the attic will be cold and the people in the ramshackle wing may well freeze to death. Those people will have a real incentive to take control of the heating system, even if it means going to war with the rich people. On the other hand, if you crank the thermostat high enough to keep the people in the basement and the attic warm, the well-to-do folks may get overheated. And the people in the ramshackle wing will undoubtedly demand that the wealthy residents pay to install a heating system in their part of the house.

Together, everybody might attempt to set up committees and governance structures to try to work this out in a way that is fair to rich and poor alike. They might wring their hands about the complex moral choices they have to make. But when push comes to shove, the rich folks are likely to do whatever is necessary to ensure that they are comfortable, whether it means paying to have the heating system reconfigured or hiring thugs to guard the thermostat. That's simply the way these things work. In a big old house like this, there are always going to be winners and losers.

A geoengineered world would be a little like this. In the distant future, we may understand the positive and negative feedbacks that modulate the earth's climate well enough to develop what Ken Caldeira calls "a tailored climate." In other words, you tell climate scientists what temperature and rainfall you want, and they tell you how much sunlight you need to shield—and where and how—to achieve your goals. But that world is still a long way off. For now, it is likely that the rich, privileged, technologically sophisticated nations with the capacity to build and operate a geoengineering system will be calling the shots and setting up the governing structures, not the poor, needy, undeveloped ones.

According to Scott Barrett, a professor of natural resource economics at Columbia University who has done a lot of thinking about geoengineering governance, one rough analogue might be the Large Hadron Collider, which was recently built near Geneva, Switzerland, to test the standard model of particle physics. The knowledge gained from this $9 billion project will benefit the global public good, but until it was turned on, it was impossible to rule out a scenario in which the experiment could create something called a strangelet—a hypothetical particle that could start a chain reaction, transforming ordinary matter into what scientists call "strange matter," and that could, as one scientist put it, "transform the entire planet Earth into an inert hyperdense sphere about one hundred meters across." Scientists involved in the project believed that the chances of this happening were about the same as the chances that alien spaceships would land on the White House lawn. But since no one had ever done the experiment before, no one really knew for sure. "Existing theories [about the safety of the collider] are reassuring, but they have not been tested," Barrett wrote in 2008. "And do we really want to test them? Are we sure that the global public good of new knowledge outweighs the global public bad of the risk of annihilation? More importantly, who should decide whether the experiments should go ahead?"

In the case of the collider, the countries with money and power have called all the shots: the twenty European members of CERN (officially, the European Organization for Nuclear Research) and its partners in this project—India, Japan, Russia, and the United States. No one asked the president of Kenya—or the prime minister of Australia, for that matter—how he felt about the risk of this experiment. Nobody asked me or my friends if we thought this was a reasonable risk, and I'm sure nobody asked anyone in Ethiopia about it.

There are other ways of thinking about collective agreements, of course. As an example, Barrett points to the debate about whether to destroy the remaining stocks of smallpox virus. Smallpox was eradicated in 1977, which was a very good thing not only for the countries where the disease raged (such as India) but for humanity in general. In the twentieth century alone, this fearsome disease killed 300 million to 500 million people. But just because the virus has been eradicated doesn't mean the threat is gone. If smallpox were somehow reintroduced today, either through an accidental release or a bioterrorist attack, it would be more deadly than ever (in part because public health officials have stopped vaccinating people against it). After 1977, fears about the security of the virus compelled laboratories around the world to destroy or transfer their stocks. By 1983, known stocks were held by just two World Health Organization "collaborating centers," one in Atlanta and the other in Moscow. The question was, what should be done with these few remaining stocks?

"Again the question: Who should decide?" Barrett wrote. "The two states that possess the virus obviously have the upper hand (just as the major powers would have the upper hand in developing a geoengineering project), but being WHO collaborating centers, the labs in Atlanta and Moscow are obligated to serve the global interest."

The WHO polled its 190 members: should the last known stocks of the virus be retained or destroyed? Russia wanted to hold on to its stocks; Britain, France, Italy, and the United States were undecided; every other country favored destruction. (Fearing a bioterrorist attack, the United States later decided that it wanted to keep its stocks after all.) To settle the matter, the WHO worked out a compromise, passing a resolution that called for the virus to eventually be destroyed, but allowing the United States and Russia to keep the virus for research purposes for an indefinite period of time.

As Barrett points out, the arrangements surrounding the decision to keep smallpox are very different from those connected with the conduct of possibly dangerous experiments at the Large Hadron Collider:

The latter are being undertaken by a relatively small number of countries, without wider consultation let alone approval. The smallpox decision, by contrast, has been undertaken in a setting in which all the world's countries were invited to take part. To be sure, in this case the power relations among countries are vastly unequal. But the process that emerged favored consensus — an especially fortunate outcome. Since every country will be affected by whatever is decided, it is as well that each should agree with the decision.

It's not hard to guess how a similar attempt to forge an international agreement might play out in a geoengineering context. Some yet unborn branch of the UN, or some standalone group — let's call it the International Atmospheric Trust — could be charged with making decisions about regulating the earth's climate. How much of the Arctic do we allow to melt? Should we focus our attention on alleviating drought in the Sahel or restoring the Asian monsoon? Should we preserve the climate as it was in, say, 1990? What should we do if a majority of people actually decide they *like* a warmer world?

If all this sounds like notes from an environmental ethics seminar, it shouldn't. The truth is, we are *already* making these kinds of choices. We are just doing it quietly. When we (or the politicians we elect) choose targets today for CO_2 reductions in 2020 or 2050, we are making an implicit judgment about what kind of climate we want and what kind of climate we are going to leave to our kids.

But in a geoengineered world, the political dynamics will be very different. These judgments about what kind of climate we want will become an explicit part of the conversation — and that will change everything. Once the idea of intentionality is injected into the process, so are issues of fairness and social justice. As writer and former Episcopal priest Garret Keizer put it in a recent essay called "Climate, Class, and Claptrap," "You do not repair the climate for an entire planet without staggering sacrifices, and people will not elect to make staggering sacrifices unless the burden is shared with something like parity." And with intentionality comes something else: lawsuits. Indeed, it is a running joke among scientists that the operational costs of tweaking the earth's climate will be nothing compared to the legal costs of these schemes. "Every rainy day, every early frost, every drought will suddenly become someone's fault," one scientist speculated. "People will file lawsuits if their corn crop dies or they crash their car in an ice storm."

But there's one fact in all this that is beyond dispute: global warming will hit the poor the hardest. "The poor have always suffered from bad weather," Bill Gates told me one morning as we talked in his office in Seattle. He was dressed in a blue polo shirt and loafers; outside the window, sailboats bobbed in the drizzle. "We don't suffer from bad weather. I sit here and watch it go by, but I am air-conditioned. If you're up here in [the] temperate zones, the additional CO_2 is probably a net win for crops and stuff. It's mostly the poor areas where the high variability of the weather is going to be a problem, because you don't have irrigation, and you don't have buffer food stocks and that sort of thing."

In Gates's view, traditional ideas about how the rich might help the poor confront global warming are inadequate. "People say, 'Let's do climate adaptation money for the poor, then we'll avoid their having a problem,'" Gates told me. "Well, they already have a problem. Millions of them die now, and the things you put the money into, in terms of seeds or water storage, are actually the same things that are being done now but should be done even more." Gates is a big believer in clean energy research and development, but he also fears that the push for clean energy might suck up money that could be better spent in other ways. "You could say, 'Let's take the money spent on vaccines and help the Chinese build another energy structure.' You could use up all the foreign aid easily just by paying the Chinese to do something different."

How might geoengineering change this equation? Let's say that parking cloud-brightening machines off the coast of Africa could increase rainfall in the Sahel, or injecting particles into the atmosphere could help slow a sea level rise that would be devastating to millions of people in Bangladesh. If the West could develop technologies that would have a reasonable likelihood of reducing climate impacts in the developing world, would it be obliged to use them? Instead of viewing geoengineering as a scheme to increase the comforts of the rich, perhaps we could just as easily see it as a way to reduce the suffering of the poor—especially if climate impacts continue to accelerate. Gates, to his credit, understands this as well as anyone. "There is a lot of basic science that needs to be funded and investigated here," Gates told me. "And there's a good chance we may choose not to deploy geoengineering. But if, for example, we start to see really dramatic ice melt in Greenland and rising sea levels, then we'd absolutely have to look at geoengineering to try to slow things down."

In any conversation about geoengineering governance, there is an elephant in the room that nobody knows how to handle: the military. It's not hard to understand why men in camouflage uniforms give scientists and politicians the jitters. For one thing, it is a reminder of how easily many geoengineering schemes could be transformed from a tool to save civilization to a tool to dominate civilization. Indeed, few subjects bring conspiracy theorists out of the woodwork faster than the notion that a shadow arm of the U.S. government is in cahoots with the military to engineer the climate. And as I mentioned earlier, there is a whole lunatic fringe of sky watchers who believe that the skies are already being sprayed with heavy metal–laced particles as part of a covert depopulation and mind control agenda. "Governments exist to govern, to exert power, to achieve control, to dominate the many for the benefit of the few," one chemtrails website explains. "Chemtrails provide one of many, brilliant yet still-to-be-perfected state systems to control the masses."

Crazy? Yes. But as the poet Delmore Schwartz once said, "Just because you're paranoid, it doesn't mean no one is following you." In fact, it's not easy to see how a serious geoengineering program could move forward without some degree of military involvement, both here in the United States and in countries such as China and Russia. This is particularly true in the case of stratospheric aerosols. If you want to inject particles into the stratosphere, you need airplanes, high-altitude balloons, or big artillery guns to get them up there. Even if you just want to run a large-scale field experiment, aircraft and sophisticated monitoring instruments are necessary. Yes, one can imagine scenarios in which this could be done without the military, especially in the United States, where NASA could well play a major role. But the military has many strengths that make it a likely candidate to run an operation like this, including access to satellites for observation, access to congressional budgets for funding, and experience in carrying out large, complex operations in the oceans and skies. It's easy to envision the military responding to warnings of a catastrophic methane release in the Arctic the same way it would respond to a humanitarian crisis in central Africa or a natural disaster in Southeast Asia.

So here's a key question: is it possible to have an open, transparent, publicly accountable geoengineering research program that includes the military?

History is not encouraging. Since the days of the Roman Empire, weather control has been a fantasy of military generals. During the Cold War, many of the same scientists who worked on the nuclear bomb were also interested in climate and weather modification. The emergence of computers in the late 1940s only deepened that connection. John von Neumann, a Hungarian-born mathematician who worked at Los Alamos but is better known as the godfather of modern computer programming, believed that computers would someday allow scientists to tackle problems they had never imagined they could solve. Chief among them, in von Neumann's mind, was controlling the weather — or, as he put it, "jiggling the planet." Von Neumann believed that with enough computing power, the physical properties of the weather could be broken down, reduced, simplified, calculated, and, ultimately, tweaked. He also grasped, however, the dangers lurking behind this endeavor. "Present awful possibilities of nuclear warfare may give way to others even more awful," von Neumann wrote shortly before his death in 1957. "After global climate control becomes possible, perhaps all our present involvements will seem simple. We should not deceive ourselves: once such possibilities become actual, they will be exploited."

But it was weather modification on a small scale, rather than climate control on a larger scale, that really captured the interest of the military. Cloud seeding with silver iodide — the technique developed by Irving Langmuir and Vincent Schaefer at GE during the 1940s — was particularly attractive as a potential battlefield weapon. Between 1967 and 1972, weather modification took what historian James Fleming has called "a macro-pathological turn" in the jungles of Vietnam. Project Popeye, as the scheme was called, was a secret cloud-seeding operation conducted to reduce foot traffic along portions of the Ho Chi Minh Trail. It didn't work. For five years, the U.S. Air Force flew thousands of cloud-seeding missions out of an air base in Thailand without the knowledge of the Thai government, but with the full and enthusiastic support of President Lyndon Johnson. Then in 1971, columnist Jack Anderson broke the story about Air Force rainmakers in Southeast Asia in the *Washington Post.* By 1973, the Senate had adopted a resolution "to prohibit and prevent, at any place, any environmental or geophysical modification activity as a weapon of war."

Project Popeye was only one of a number of weather modification projects taken up by the U.S. military during the Cold War. There was also Project Stormfury, which tried to modify hurricanes by cloud seeding (Fidel Castro once accused American scientists of using hurricanes as a counterrevolutionary instrument of war), as well as Telleresque dreams of using nuclear explosions to change the weather. But by the 1970s, the military's interest in weather modification was largely dead. "As a tool of warfare," Lowell Wood told me, "weather modification was a total failure."

Still, the old fantasies kept popping up. In 1996, a small group of officers in the U.S. Air Force issued a report called "The Weather as a Force Multiplier: Owning the Weather in 2025." The report claimed that "in 2025, U.S. aerospace forces can 'own the weather' by capitalizing on emerging technologies and focusing development of those technologies to war-fighting applications." In addition to conventional cloud-seeding methods, the Air Force visionaries proposed computer hacking to disrupt an enemy's weather monitors and models and the creation of clouds of particles that could block an enemy's optical sensors. The report pointed out that weather modification, unlike other approaches, "makes what are otherwise the results of deliberate actions appear to be the consequences of natural weather phenomena."

To some, the report was evidence of the inevitability of military involvement in geoengineering projects. "Given such mindsets," Fleming wrote of the report, "it is virtually impossible to imagine governments resisting the temptation to explore military uses of any potentially climate-altering technology." But in my view, this report, now almost fifteen years old, shows, if anything, the cluelessness of the military — or at least the writers of this report. If you read it closely, the report demonstrates no new thinking and little understanding of the complexity of the science behind "climate-altering technology." It's really just old ideas about cloud seeding dressed up in high-tech language.

The main concern for the military today is not weather modification, but the national security implications of climate change. "Recent war games and intelligence studies conclude that over the next 20 to 30 years, vulnerable regions, particularly sub-Saharan Africa, the Middle East and South and Southeast Asia, will face the prospect of food shortages, water crises and catastrophic flooding driven by climate change that could demand an American humanitarian relief or military response," the *New York Times* reported in 2009. A recent military exercise explored the potential impact of a destructive flood in Bangladesh that would send hundreds of thousands of refugees streaming into neighboring India, touching off religious conflict, the spread of contagious diseases, and vast damage to infrastructure. "It gets real complicated real quickly," said Amanda J. Dory, the deputy assistant secretary of defense for strategy, who is working with a Pentagon group assigned to incorporate climate change into national security strategy planning. Or as Navy vice admiral Dennis McGinn told a recent gathering of environmental journalists, "Climate change is going to be a threat multiplier in unstable regions across the world."

One sign that the military is taking these issues seriously was evident in March 2009, when the Defense Advanced Research Projects Agency (DARPA) held a daylong workshop at Stanford University on the prospects for geoengineering. DARPA has already given birth to a number of transformational technologies, including the Predator drones commonly used in the Iraq War, as well as the Internet itself. It's the agency most likely to pursue research on geoengineering hardware—sprayers for stratospheric particles, for example, or pipelines to high-altitude balloons, or even hydrogen guns.

The mere fact that DARPA was nosing around made some scientists uneasy. "The last thing we need is to have DARPA developing climate-intervention technology," Caldeira told me. He agreed to go to the meeting, "to try to get DARPA *not* to develop geoengineering techniques. Geoengineering is already so fraught with social, geopolitical, economic, and ethical issues—now we're going to add the military?"

But Gregory Benford, the astrophysicist who suggested the Arctic experiment to Wood and Caldeira (see chapter 6), believes that the only way to get serious research done on geoengineering is with military involvement. "The military can muster resources, and they don't have to sit in front of Congress and answer questions about every dime of their money," he told me. In other words, military involvement is one way to get around the need for public accountability during the early research phase. That might be helpful to impatient scientists like Benford in the short term, but in the long run this approach will only breed suspicion and paranoia about what scientists are really up to. The only way to build public confidence in geoengineering is to invite public scrutiny. In this context, the military is useful only because it has a lot money and a lot of sophisticated hardware that might be directed toward geoengineering research.

At the DARPA meeting, Benford made the case for injecting stratospheric aerosols above the Arctic, arguing that it is the most vulnerable place on the planet, due to the region's rapidly rising temperatures and melting ice, as well as the best place to experiment with aerosols, due to the region's atmospheric conditions and sparse population. Benford, who has been an adviser to both DARPA and the CIA for thirty years, was invited back a few months later to speak in more detail at the agency's annual board meeting at the Chaminade Resort in Santa Cruz, California. At the second meeting, Benford explained that the military could inject the particles with KC-10 Extenders, heavy aircraft used for refueling fighters but also capable of high-altitude cruising. Benford estimated that the experiment would cost about $200 million but that the wealth of scientific data would be "incalculable."

In Benford's view, there are two reasons why the military might be interested in cooling off the Arctic. "First, they are worried about the melting ice in the Arctic, because they use the sea ice for scrambling the signals for submarines," he told me. "They go under the ice in one place, then out in another, playing peekaboo with satellites. If they lose the Arctic ice, they lose that capability. And second, the thawing tundra in the region is leading to the possibility of a large methane release in the region. If that happens, it could be game over for all of us. Cooling the Arctic off with particles in the stratosphere may be the only way to stop it." Benford didn't mention it, but it's also true that a rapidly melting Arctic may touch off a new Great Game in the region, with various countries fighting for rights to new shipping routes and fossil fuel deposits. If and when that happens, the military is likely to become involved in the Arctic whether it likes it or not.

In the United States, the military's initial involvement in geoengineering may consist of nothing more than monitoring and verifying field studies. But once the cannonball gets rolling, who knows where it will end up? It could be that, like cloud seeding, the various technologies never work well enough, or precisely enough, to interest the military. Or it could be that the military's interest will be purely defensive—that is, driven by the need to better understand what kind of trouble could be caused if the Russians decided to start experimenting with stratospheric aerosols. Or it could be something far more insidious, such as the beginnings of a Cold War–style military-industrial-climate complex.

As it happened, David Keith and I spent a lot of time talking about the Cold War and the lessons it holds for geoengineering. Keith is a strong believer that the military is *not* the right agency to run a geoengineering program, arguing that it would be much better handled by NASA, which has a lot of experience in similar research-intensive operations, and the National Oceanic and Atmospheric Administration, which has a lot of experience in planetary-scale monitoring and observation. He points to a program NASA developed in the 1990s to study the atmospheric effects of supersonic aircraft as one model for how geoengineering research and field-testing might begin. "If we were ever to do this, it needs to be an open and collaborative undertaking," Keith told me. "NASA is far better at that than the military."

One of the reasons that Keith is deeply committed to openness and transparency is that he sees himself as part of a reaction to Cold War culture. In fact, a good way to make Keith nervous is to ask him—as I did one night in his living room—if he worries that some of the geoengineering ideas he is working on might someday be co-opted by the military and used for purposes he never imagined. That was, after all, what happened with the atomic bomb. Many of the scientists, including J. Robert Oppenheimer, believed that they were great patriots and heroes while they were working at Los Alamos. Later, after more than 200,000 people died at Hiroshima and Nagasaki and an international arms race had begun, many of these same scientists felt differently.

"I think about it a lot," Keith said. Then he ran upstairs to get his laptop. When he returned, he opened it and searched though his files. "Rob Socolow just sent me something that is exactly about this," he said. Socolow is a professor of mechanical and aerospace engineering at Princeton and a longtime colleague of Keith's. Like

Keith, he has given the moral dimension of geoengineering a lot of thought.

"Here it is!" Keith exclaimed. It was an excerpt from the memoirs of Andrei Sakharov, the Russian nuclear scientist who later turned antinuclear activist. Keith explained that the excerpt was an account of a banquet of scientists and military officers after a successful nuclear test in Russia in the 1950s. The chief military officer at the table was Marshal Nedelin, whom Sakharov described as "a thick-set, stocky man who spoke softly but with a confidence that brooked no objection."

Keith began to read aloud, his laptop balanced on his knees:

When we were all in place, the brandy was poured. The bodyguards stood along the wall. Nedelin nodded to me, inviting me to propose the first toast. Glass in hand, I rose and said something like: "May all our devices explode as successfully as today's, but always over test sites and never over cities."

The table fell silent, as if I had said something indecent. Nedelin grinned a bit crookedly. Then he rose, glass in hand, and said: "Let me tell a parable. An old man wearing only a shirt was praying before an icon. 'Guide me, harden me. Guide me, harden me.' His wife, who was lying on the stove, said: 'Just pray to be hard, old man. I can guide it in myself.' Let's drink to getting hard."

Keith laughed nervously. "Isn't that great!"

I laughed, too, but only for a moment. "So this is your fear — you and other scientists work on technology to geoengineer the climate, and you feel okay about it because you know you are ethical people, and you know you are doing it for the right reasons. But then along comes the military, or someone else, and they have other ideas about what to do with this new technology you've developed."

"Exactly," Keith said.

"Do you think that will happen?"

"I have no idea," he said, closing his laptop. He was no longer laughing.

....

TEN

Human Nature

NOBODY LIKES TO ADMIT IT, but most of us have very little connection with the natural world anymore. We breathe the air; we admire (or curse) the sky; we sweep up the leaves when they fall on
our lawn. And, of course, if a major weather event strikes somewhere in the world — a hurricane, a flood, a tsunami — we watch
footage of it on TV, bingeing on disaster porn. But in our day-to-
day lives, we hardly think about it. Our houses and cars are air-conditioned; most of us work in jobs that have little or no connection
to the weather outside; we entertain ourselves by watching pixels
dance in various patterns on digital boxes. In fact, you could make
the argument that the whole purpose of technology — starting with
the invention of fire — has been to distance ourselves from the wan-
ton ways of Mother Nature.

Of course, there are billions of people alive today who are not
very distant from nature at all, because they are sleeping on the
streets in Los Angeles or living in a hut in the Ethiopian desert or
the Burmese jungle. And it is true that images such as the famous
"blue marble" photograph of the earth taken from *Apollo 17* have
inspired millions of people to think differently about the planet we

live on. But in general, it's hard to deny that in our daily lives, technology has taken us farther away from nature, not drawn us closer to it. And if you accept the argument that geoengineering is likely to be attempted by rich, technologically sophisticated nations, then you also have to accept that it will be carried out by people whose lives are increasingly remote from the world they want to manipulate. I don't want to say that the idea of monkeying around with the climate is like a big video game to us, but sometimes it does feel that way.

Technology is not the only thing that distances us from the natural world. Religion can, too. It goes without saying that many religious texts are full of respect and reverence for the natural world. But faith in a higher power can also be an excuse to disregard the importance of our relationship with nature and lead to a willful ignorance of the laws of physics. The other day, a friend emailed me a YouTube clip of Don Blankenship, the CEO of Massey Energy, the biggest and most ruthless coal company in Appalachia, speaking at a pro-coal rally in West Virginia. Like many old-fashioned coal barons, Blankenship thinks the idea that human beings have anything to do with global warming is a myth propagated by tree-huggers and anticapitalists. A few minutes into his speech, he said, "Only God can change the earth's temperature, not Al Gore."

This is not an uncommon view. Despite all our high-tech weather monitoring equipment and sophisticated understanding of the climate system, the sky is often still viewed as the realm of God, just as it was three thousand years ago when Zeus was kicking up storms with his temper tantrums. A year after Hurricane Katrina devastated New Orleans, Pastor John Hagee, a well-known Evangelical Christian figure who preaches at a megachurch in Texas, said that God was punishing the city for its decadent ways and especially for hosting a gay parade. "All hurricanes are acts of God," he pronounced, "because God controls the heavens." Indeed, conservative religious leaders are likely to see geoengineering as a sacrilege against God—or, at the very least, a usurping of his or her powers.

"If you really do go to geoengineering, essentially the impression is going to be that there's no further acts of God with respect to natural events," Mike MacCracken, a respected climate scientist, told me. "It's something you have to approach very cautiously."

Maybe you do. But my problem with crediting God with every hurricane and rainstorm, as well as every sunny day, has nothing to do with the fact of his or her existence (or lack thereof). I come from a very long line of Methodist ministers, and I'm certainly willing to acknowledge that they knew something about the universe that I don't. My problem with this view has to do with responsibility. If you believe that God is in charge, it's easy to sit back and let him do the driving.

Oddly enough, many people who think of themselves as environmentalists are just as passive before nature as an Evangelical preacher. They see nature as something to be revered, worshiped, preserved. This is a very American perspective, one that was most fully expressed by the founding of national parks in the late nineteenth century. I'm a big fan of national parks and believe that the preservation of places such as Yellowstone and Yosemite was an act of profound foresight. But national parks have also encouraged a view of nature as a place that is cut off from the rest of the world, a place where you carry out all your trash and you still feel as if you're trespassing on hallowed ground. And in a way, you are. The whole point of a national park is to capture nature under glass in its most glorious and benevolent mood. That's all well and good, except that given the state of the world right now, our future may depend on our embracing a radically different view of our relationship with nature.

Don't get me wrong. I hope that we never launch particles into the stratosphere, dump iron into the oceans, or brighten clouds. I hope that we will grasp the scale of the catastrophe that awaits us, muster up the courage and political will to cut emissions quickly and deeply, invent new energy systems that are cheap and clean and abundant. Most of all, I hope that the whole notion of geoengineering looks in retrospect exactly how it looks at first glance: like a bad sci-fi novel writ large. But whatever we end up doing, it is time to discard our romantic fantasies about Mother Nature and realize that she is not a benevolent force—she is, as Lee Silver, a molecular biologist at Princeton University, has put it, "a nasty bitch" who is capable of wiping out billions of people without shedding a tear.

Geoengineering is not about preserving nature, but about working with it, bending it to our will. One of the fears about engaging in deliberate manipulation of the climate is that there will be some kind of deeper spiritual loss accompanying this effort. If we take an active role in managing the planet, we will replace mystery and beauty with lousy engineering. If it rains too little, we will curse the engineers at Climate Control Central. If it rains too much, we will curse them again. Once in a while, they will get it right, and once in a while, we will be grateful. In the not-too-distant future, you can imagine the International Atmospheric Trust hosting a banquet each year for the Planetary Engineer of the Year Award. When you think of it this way, geoengineering does indeed sound like a very bad idea.

But I want to talk about this idea that the thrust of the human hand into a new realm will debase it. That is a feeling I know something about, because it is precisely what I felt more than a decade ago when my wife and I decided that she would undergo in vitro fertilization. We both wanted a child, and if we wanted a child, this was likely to be our only option. But I did not go into it easily. Life is supposed to be created in the womb, not a petri dish. It felt like a mechanical intrusion into a private realm, a diminishment of something sacred.

In reality, I found the opposite to be true. The more that was revealed to me about how life is created, the more deeply I appreciated it. I was particularly moved by a photograph that our doctor gave us of the embryos when they were a few days old. The picture, which was taken through a powerful microscope, showed seven embryos, each of them a small, rough clump of cells. It was impossible to imagine that these clumps contained everything necessary to create a human being — that they were, in fact, the germinating seeds of two children I love today (we had twins). I felt a similar way a month or so later when my wife's obstetrician put an instrument over her abdomen and I heard the sound of two freight trains rolling down the tracks. Heartbeats! Two of them! How did that happen? It makes me think now of a line I came across in Charles Darwin's autobiography: "The more we know of the fixed laws of nature, the more incredible do miracles become."

I've discovered that the people who understand this best are gardeners. I'm not much of a gardener myself, but I am married to one. My wife, Michele, is happiest when she has dirt under her fingernails, and one of her highest aspirations in life is to grow all our own food. It's because of her that we have chickens in an urban backyard, and it is because of her that our kids have such a heightened sensitivity to the freshness of green beans that they can take one bite and tell you, with a good chance of being correct, whether the bean is store-bought or homegrown.

My wife's garden is, by any standard, a product of human artifice. There is nothing "wild" about it, nothing undisturbed, nothing left alone. She has planted every plant and mixed the soil to her liking with imported alpaca manure. The garden is entirely organic — she's no more likely to use Miracle-Gro than she is to dye her hair pink — but it is also entirely human. It is an artifact, but it is a living artifact. You do not walk through her vegetable garden and admire the basil and the asparagus and feel that nature has been banished. Quite the contrary. You feel how well she has been able to work with nature to create this miraculous little patch of ground that both feeds us and is an object of a particular kind of beauty. It is not the same feeling you get when you hike through the California redwoods or visit the Grand Canyon. It's not Mother Nature hitting you in the forehead with her majesty and grandeur. But in its own modest way, it is equally powerful. On their own, neither my wife nor Mother Nature could create anything like this. But by collaborating, they have achieved something new, and something that brings out the best in both of them. In my wife, it has inspired a deep feeling for this otherwise unremarkable piece of ground in upstate New York — a feeling that she would describe as something like love.

Could we do the same thing with the planet? Obviously not in the same way. We're never going to control the earth the way my wife controls her garden. But it does point to a different way of thinking about our relationship with the world we live in and why getting into the business of actively managing the climate might not be such a bad idea after all, especially given the dire circumstances we find ourselves in. In a March 2009 speech about the challenge of global warming, Rowan Williams, the archbishop of Canterbury, argued that the greatest danger we face is not technological hubris, but human apathy: "In the doomsday scenarios we are so often invited to contemplate, the ultimate tragedy is that a material world capable of being a manifestation in human hands of divine love is left to itself, as humanity is gradually choked, drowned or starved by its own stupidity."

I do believe this is what it comes down to. We can use our imagination and ingenuity to create something beautiful and sustainable, or we can destroy ourselves with stupidity and greed. It is our choice. Geoengineering may well turn out to be yet another tool of dominance, a newfangled way for human beings to screw things up even faster. But it doesn't have to be that way.

As David Keith said to me as we were riding up a ski lift one day, "We're in the gardening business now, damn it!"

In the end, it's important to underscore the fact that the rising interest in geoengineering is driven less by mad scientists than by spineless politicians. The failure of the Copenhagen climate conference in December 2009 is the latest case in point. After a two-week-long meeting that was marked by political tensions between the rich nations that are dumping the greenhouse gases into the atmosphere and the poor nations that are feeling the impacts of the rising heat, world leaders emerged with a weak political agreement to do little more than monitor their emissions in the future. Despite the scale and urgency of the threats we face from climate change, there was no legally binding commitment to cut greenhouse gas pollution. The *Financial Times* called it "the emptiest deal one could imagine, short of a fistfight." The Royal Society, Britain's most respected scientific body, said that the world was now "one step closer to a humanitarian crisis."

However you feel about the deal that was brokered in Copenhagen, it's pretty obvious that we are not going to make serious cuts in emissions anytime soon and that it is now time to prepare in a practical-minded way for life in a rapidly changing climate. That means exploring a wide variety of adaptive measures to increase the resiliency of our cities and towns, from building dikes against the rising seas to securing more reliable drinking water supplies, as well as considering how we will cope with the suffering and turmoil that a changing climate will bring to the developing world. And it means acknowledging the fact that geoengineering is a tool that might help reduce the risk of catastrophe.

A good example of how the conversation is already shifting is the sudden enthusiasm for white roofs. The idea, first promoted by California energy commissioner Arthur Rosenfeld, who has long been the guru of energy efficiency in my home state, is a no-brainer: by reflecting sunlight, white roofs lower cooling costs in hot climates. In urban areas, they also reduce the heat-island effect—the tendency of cities, with all their built infrastructure, to absorb warmth like a big pile of rocks. The idea has quickly been taken up by politicians around the country as a low-cost, low-impact way of both increasing building efficiency and increasing street cred with green-minded voters. But what is really important about white roofs is that the idea has injected the word "albedo" into energy and climate discussions. And that opens the door to thinking differently about how to solve our energy and climate problems. Ultimately, no matter how many roofs we paint white, it's not going to have a big impact on the earth's climate. But it might have a big impact on our imaginations. Because once we grasp the idea of white roofs, other albedo-changing ideas, such as particles in the stratosphere and cloud brightening, make a lot more sense.

So how soon might we get serious about geoengineering? Well, that depends on how temperate and rational you think we are. The moral and ethical taboos regarding geoengineering seem to be fading fast. But given all the basic research that still needs to be done, not to mention the observational systems that need to be in place and the hardware that needs to be completed, it's hard to see how it would take less than a decade or two to get anything going beyond large-scale field tests. And that would be a crash program. More likely, this is going to evolve over the next thirty years or more. But who knows? If we find ourselves in the middle of the climate equivalent of the subprime mortgage meltdown, there is no telling how soon we might start throwing dust up into the sky.

And that is why it's imperative that we learn as much as we can as quickly as we can about the strengths and weaknesses of the various technologies I've discussed in this book. The question, of course, is how best to do that. When it comes to pulling carbon out of the atmosphere, for example, market forces, combined with the carbon market that is evolving out of Kyoto and Copenhagen, are likely to separate the wheat from the chaff. And because everyone agrees that carbon will soon be a valuable commodity, it's easy to see how building a CO_2 extraction device, such as David Keith and others are attempting, could be a profitable enterprise. In fact, the jerry-rigged contraption I saw on my first visit with Keith in 2006 has evolved considerably, and he has started his own company, Carbon Engineering, to push the technology even further. Within just a few months, he was able to attract more than $3 million in investment from Calgary businessmen and Bill Gates. And Keith is not the only one. Klaus Lackner, a physicist at Columbia University who, like Keith, is a pioneer in the field, is trying to commercialize air-capture devices by using an entirely different technology. Peter

Eisenberger, another Columbia physicist, is also starting a company to develop air-capture technology. All this is driven, of course, by money. Carbon engineering is essentially a game of economics.

Albedo engineering is far more complex. There's no equivalent of a carbon market for albedo. More important, messing with the earth's albedo is where most of the danger lies, politically and environmentally. Clearly, the best risk-reduction tool is knowledge. But how to move forward? In Ken Caldeira's view, this is a job for governments, not entrepreneurs. He believes it's time to push for what he calls a "Climate Emergency Response Program"—that is, a federally funded research program that looks at what the options might be if we had to cool off the planet in a hurry. When I asked him how this program might be structured, he emailed me a list of his top priorities, which included a focus on scalable technologies that could be deployed rapidly, such as stratospheric aerosols; establishment of a high-quality peer-review process to address questions of scientific and technical content ("no pork barrel funding"); and a commitment to openness and transparency in all matters ("no secrets").

How much would a Climate Emergency Response Program cost? "Ten million can be spent well soon, mostly on environmental science," Caldeira wrote. Simultaneously, a National Academy of Sciences panel could begin an open collaboration with scientists in other countries to map out a longer-term research program. That's where the bigger bucks come in. Real experiments involving ships and planes and so on would cost tens of millions of dollars apiece. Then we would have to engineer prototypes of delivery systems such as cloud-brightening machines or aerosol sprayers. A scaled-up program, Caldeira believes, could cost in the $100 million per year range. "If you ever got serious about wanting to deploy, then you would need satellite observing systems monitoring Earth's radiation budget and climate response, and this could get in the billion dollar per year range," he wrote. "I would think a fully deployed system, would end up being in the $30 to $50 billion per year range, with the resources evenly split between monitoring systems and deployment hardware." This is, of course, for a global system. How these costs might be distributed is unknown, but it is likely that the United States — as well as the European Union and possibly China and India — would end up paying for most of it.

Not that the numbers really matter. "I think we won't deploy unless we are in deep trouble — and then we won't worry about the financial cost of deployment," Caldeira wrote. He pointed out that the United States spent has spent more than $700 billion on the Iraq War — nearly $90 billion in 2009 alone. The U.S. Treasury spent $2 trillion to bail out banks in 2008–2009, with no end in sight. "We wouldn't deploy these emergency climate intervention systems unless climate change was many times more damaging than the Iraq war," Caldeira wrote. "In such a context, $10 billion vs. $100 billion is irrelevant."

Two years ago, I would have argued that there was no way a Climate Emergency Response Program would be funded by Congress anytime soon. Now, post-Copenhagen, I'm not so sure. Politically, one of the virtues of such a program would be that it wouldn't ask for any sacrifice from anyone, least of all big polluters. It doesn't take much imagination to see how Big Oil and Big Coal could use the prospect of geoengineering as yet another way to divert our attention away from the need for deep cuts in greenhouse gas pollution. In fact, you could say that supporting geoengineering research is an easy way for both industry and politicians to look serious about the climate crisis while at the same time continuing to kick the can down the road. And it may be. But it is also true that a well-designed research program could open the door to a new era of scientific knowledge, one that not only improves our understanding of the risks and benefits of geoengineering schemes but also greatly extends our understanding of how the planet works.

So what do I fear? I fear that geoengineering will be packaged and sold as a quick fix. Rather than engaging people in the act of managing the planet, it will be used as another tool to increase our passivity — just sit back and let Big Brother take care of the climate! Even people who should know better fall into this trap. "If we could come up with a geoengineering answer to [global warming], then Copenhagen wouldn't be necessary," Richard Branson told the *Wall Street Journal* in October 2009. "We could carry on flying our planes and driving our cars." Wrong! *Geoengineering is not a substitute for cutting emissions.* It is not a substitute for massive spending on clean energy research or for political action. It is not a substitute for changing everything about our lives — where we live, how we make and consume energy, and how we feed and clothe ourselves. In the short term, geoengineering might be a way of buying us more time or forestalling a climate catastrophe. But it is not a "get out of jail free" card.

In fact, if we lived in a rational world, instead of diminishing the political will to reinvent our energy economy, the prospect of geoengineering would alarm us enough to boost it. "Thinking seriously about geoengineering should prompt more people to understand the ways in which we are already manipulating the climate," said Dan Dudek, chief economist at the Environmental Defense Fund, one of America's most influential environmental groups. "It might also serve as a warning: this is absolutely where we are headed if we don't get our act together and cut emissions."

Another fear I have is that well-meaning environmentalists and do-gooders will stall funding on geoengineering research for another decade. This would be dangerous and counterproductive. The most important thing we need right now is to better understand the risks and benefits of various geoengineering schemes. During the three years I spent writing and reporting this book, geoengineering existed largely outside the miasma of lobbyists, Beltway environmentalists, energy industry fat cats, regulators, bureaucrats, and fearmongers who stink up most debates about energy and climate issues. But as that changes, the conflict over geoengineering will become coarser and more politicized. It will be very easy for the media to crank up fears about crazy scientists, especially given the astounding levels of scientific illiteracy in America. The spin will be about hubris, not risk reduction. It will be about ideology, not practicality. We may well learn that the dangers of geoengineering far outweigh the benefits and that even in the event of a climate catastrophe, we're better off riding it out than trying to cool the planet with particles in the stratosphere. But unless we do the research, we won't know for sure.

Caldeira, as usual, put it well. "The way I look at it is that we're talking here about people's lives, and I don't think we're going to deploy these systems to save polar bears," he told me in one late-night conversation. "I think if they're going to be deployed, it's going to be to help people from dying of famines or something dramatic like that. And I think that these techniques have a potential to save lives and reduce suffering, and we should explore whether that's true or not. It sounds like the moral high ground to say, 'Oh, well, we should never interfere with the climate system.' But we're obviously interfering with the climate system wholesale now, and it's possible that more intelligent interference could reduce the damage from the first interference. But it could make it worse. I don't think we know, which is why we need the research."

Another thing I fear is that the scientists who work on geoengineering will develop excessive enthusiasm for their own work. If and when geoengineering is taken seriously as a possible response to global warming, it will have a transformative effect on many sciences, not unlike the way the development of the atom bomb temporarily pulled everything into the nuclear vortex. Obviously, science follows the money, so as funding for research projects grows, so too will the number of scientists working on them. But my real worry is not financial or institutional; it is psychological. It is the idea that the scientists who are working on this might start thinking of themselves as superheroes. That would not be good. Modesty and humility are necessary virtues for future geoengineers. As J. Robert Oppenheimer's former physics professor in Germany, Max Born, said in his memoirs, "It is satisfying to have had such clever and efficient pupils, but I wish they had shown less cleverness and more wisdom."

And yet the thing I fear most is that we won't do anything at all. We won't explore geoengineering; we won't cut greenhouse gas pollution in any significant way; we won't change our lives. We will argue about it on TV and write books and make movies and hang banners on the smokestacks of coal plants, and nothing much will change. We will just ride into the dark apocalypse that James Lovelock fears, a future of war and starvation and disease driven by the changes on our superheated planet.

If growing up in Silicon Valley taught me anything, however, it's that the future has its own agenda. And on good days, my worries fade. I can imagine a time in the not-too-distant future when my kids—grown adults by then, with children of their own—are sitting on the porch of some old house, a little like the one we live in now. Maybe the grass will be mowed by a robot, the basil in the garden will be six feet tall, and meat will grow in silver vats. Who knows what the future will bring? But I can imagine them watching a butterfly float by and bees dance in the flowers. I can imagine them feeling hope and power and beauty in all the small wonders around them. And I can imagine them looking up at the sky, noticing a silvery cloud in the shape of an elephant or a bear, and feeling thankful to the engineers who sent it their way.

Acknowledgments

Ken Caldeira, David Keith, and James Lovelock are at the top of my thank-you list. Without their trust and cooperation, this book would not exist. I'm also deeply grateful to Jann Wenner, Will Dana, and Eric Bates at *Rolling Stone,* where I have worked for more than fifteen years, for their trust, friendship, and fearlessness. Among the dozens of scientists I spoke with while researching this book, John Latham, Mike MacCracken, Ray Pierrehumbert, Phil Rasch, Alan Robock, Stephen Salter, and Lowell Wood were particularly helpful. For inspiration and insight, I relied on Kelly Wanser, Wendy Abrams, Eric Nonacs, Elaina Richardson, Dan Dudek, Armand Neukermans, David Victor, Eric Etheridge, Karen Fries, Pat Towers, Diane Cardwell, Jim Kunstler, and Carol Via. Special thanks to Martha Eddison and Tom Sieniewicz for the cocktails on the porch in Maine. My agent, Heather Schroder, was, as always, a shrewd adviser and steady pal. At Houghton Mifflin Harcourt, I'm grateful to Anton Mueller, who saw the potential in this book long before any words were written, as well as to Lori Glazer for hard work and dedication. Barbara Jatkola sharpened my dull sentences and straightened my lazy thoughts—thank you, Barbara! But the real wizard behind the curtain was Amanda Cook, a brilliant and incisive editor who also happens to be great fun to work with. Without her steady guidance and wisdom, this book would never have been completed, much less worth reading. Finally, I want to thank my wife, Michele, and our three budding CO_2 emitters, Milo, Georgia, and Grace. Their continuing faith, optimism, and love are evidence that the most powerful force in nature is not a level five hurricane; it's the human heart.

Notes

1. The Prophet

3   *Crutzen's essay:* Paul Crutzen, "Albedo Enhancement by Stratospheric Sulfur Injections: A Contribution to Resolve a Policy Dilemma?" *Climatic Change* 77, no. 3–4 (August 2006): 211–19.

3   *twenty million years:* Aradhna K. Tripati et al., "Coupling of CO_2 and Ice Sheet Stability over Major Climate Transitions of the Last 20 Million Years," *Science Express,* October 8, 2009.

5   *Mount Pinatubo:* Alan Robock, "Volcanic Eruptions and Climate," *Reviews of Geophysics* 38 (2000): 191–219.

7   *"air-condition the planet":* For early responses to Crutzen's proposal, see "Geoengineering in Vogue. . . ," RealClimate, June 28, 2006, http://www.realclimate.org/index.php/archives/2006/06/geo-engineering-in-vogue/.

7   *Coal-fired power plants:* I wrote at great length about the relationship between coal and global warming in my previous book, *Big Coal: The Dirty Secret Behind America's Energy Future* (New York: Houghton Mifflin, 2006).

7   *2007 report:* R. K. Pachauri and A. Reisinger, eds., *Synthesis Report: Contribution of Working Groups I, II and III to the Fourth Assessment Report of the Intergovernmental Panel on Climate Change* (Geneva: IPCC, 2007).

8   *15 degrees Fahrenheit:* For a good overall analysis of how global warming will transform the United States, see U.S. Global Change Research Program, *Global Climate Change Impacts in the United States* (Cambridge, Eng.: Cambridge University Press, 2009). See also Louise Gray, "Met Office: Catastrophic Climate

Change Could Happen in 50 Years," *Daily Telegraph* (London), September 27, 2009.

8 *"a different planet":* Interview with James Hansen, September 2009. Here is the full context of our conversation about sea level rise:

GOODELL: There's an emerging consensus that with a business-as-usual emissions scenario, we could see as much as one meter by 2100. Do you agree?

HANSEN: I wrote a paper called "Scientific Reticence and Sea Level Rise" because I think with business as usual, which will give you double the CO_2 by the end of the century, there's a danger of much bigger sea level changes than that ... One good paper was by Stefan Rahmstorf, who pointed out that if we assumed that the sea level rise is proportional to the temperature change, then we would get one meter [of sea level rise]. But if ice sheets begin to disintegrate, you could get much more than that. And again, look at the earth's history. The last time an ice sheet disintegrated fourteen thousand years ago, sea level went up twenty meters in about four thousand years. So that's about one meter every twenty years. So if you begin to get the dynamics taking over, [and] an ice sheet begins to collapse, the sea level rise could be greater than that. And West Antarctica is a very vulnerable ice sheet, because it's sitting on bedrock below sea level. So if we once get the dynamics started, we could see even more than the one meter level rise.

GOODELL: So you're talking about two, three, four meters?

HANSEN: Yeah. Yeah that's certainly possible. In fact, another thing that's become clear in just the last several years is that the ice shelves around Antarctica, which buttress the ice sheet, are melting at a rate of several meters per year. So that's a sign, if we continue down this business-as-usual path, that the ice sheet is potentially vulnerable.

9 *third-lowest on record:* National Snow and Ice Data Center, "Arctic Sea Ice Extent Remains Low; 2009 Sees Third-Lowest Mark" (press release, October 6, 2009).

9 *"taking the lid off":* Professor Peter Wadhams of the University of Cambridge, quoted in David Shukman, "Arctic to Be 'Ice-Free in Summer,'" BBC News, October 14, 2009.

9 *100,000 years:* David Archer et al., "Atmospheric Lifetime of Fossil Fuel Carbon Dioxide," *Annual Review of Earth and Planetary Sciences* 37 (2009): 117–34.

9 *"longer than Stonehenge":* David Archer, *The Long Thaw: How Humans Are Changing the Next 100,000 Years of Earth's Climate* (Princeton, NJ: Princeton University Press, 2008), p. 1.

10 *seventy times more powerful:* The global warming potential for methane depends on the timescale by which it is measured. According to the 2007 IPCC report, methane is seventy-two times as potent as CO_2 if measured over twenty years; it is twenty-five times as potent if measured over one hundred years. Recent studies have suggested that these numbers might be conservative. See Drew T. Shindell et al., "Improved Attribution of Climate Forcing to Emissions," *Science* 326 (October 30, 2009): 716–18.

10 *"ringiness":* Interview with Ken Caldeira, January 2009.

11 *inexplicably plunged:* Until recently, it was presumed that this sudden drop in temperature happened over a decade or so. New research, however, has found evidence that it might have occurred in a single year—or even less. See Kate Ravilious, "Mini Ice Age Took Hold of Europe in Months," *New Scientist,* November 11, 2009, p. 10.

11 *compared the earth's climate:* Wally Broecker, interview, *Morning Edition,* National Public Radio, May 12, 2004.

11 *"a fool's climate":* Interview with Paul Crutzen, August 2006.

13 *"totally insane":* Interview with Vaclav Smil, July 2008.

13 *"sign of desperation":* Interview with Michael Oppenheimer, November 2006.

13 *"Foolishness":* Interview with Kevin Trenberth, November 2006.

13 *"serious side effects":* John Holdren, "The Energy Innovation Imperative," *Innovations,* Spring 2006, p. 14.

13 *"the Frankenplanet solution":* Email correspondence with David Hawkins, November 2006.

14 *major geoengineering study:* Royal Society, "Geoengineering the Climate: Science, Governance, and Uncertainty" (report, September 2009).

14 *"worth further research":* Interview with Steven Chu, April 2009.

15 *"worth exploring":* Interview with Bill Gates, October 2009.

15 *"new climate denialism":* Alex Steffen, "Geoengineering and the New Climate Denialism," Worldchanging, April 29, 2009, http://www.worldchanging.com/archives/009784.html.

16 *"the Anthropocene":* Paul Crutzen and Eugene F. Stoemer, "The Anthropocene," *International Geosphere-Biosphere Programme Newsletter* 41 (2000): 17–18.

16 *"moderate global warming":* Royal Society, "Geoengineering the Climate," p. ix.

18 *three arguments:* For other perspectives on the troubles with geoengineering, see Alan Robock, "20 Reasons Why Geoengineering May Be a Bad Idea," *Bulletin of the Atomic Scientists,* May/June 2008, pp. 14–18. See also Gabriele C. Hegerl and Susan Solomon, "Risks of Climate Engineering," *Science* 325 (August 21, 2009): 995–96.

18 *"spin out of control":* Interview with David Battisti, November 2007.

19 *"Baptists view sex":* Steve Rayner, remarks, "Understanding the Risks of Planetary-Scale Geoengineering" conference (Lisbon, April 20–21, 2009).

20 *"Sword of Damocles problem":* Interview with Ray Pierrehumbert, October 2008.

22 *"up and running":* Interview with David Keith, September 2006.

2. A Planetary Cooler

24 *"deny them these things":* Quoted in Gregg Easterbrook, "The Man Who Defused the Population Bomb," *Wall Street Journal,* September 16, 2009.

24 *child mortality rate:* Celia W. Dugger, "Child Mortality Rate Declines Globally," *New York Times,* September 9, 2009.

29 *"significant energy requirements":* Royal Society, "Geoengineering the Climate," p. 19.

30 *"worried about him":* Interview with Deborah Gorham, January 2008.

30 *Graham Rowley:* For a full and colorful account of Rowley's adventures in the Arctic, see his *Cold Comfort: My Love Affair with the Arctic* (Montreal: McGill–Queen's University Press, 2007).

31 *"he was sleeping in":* Email correspondence with David Keith, April 2009.

33 *first presidential report:* President's Science Advisory Committee, Environmental Pollution Panel, *Restoring the Quality of Our Environment* (Washington, DC: U.S. Government Printing Office, 1965).

34 *"in laser physics":* Interview with Anthony Keith, March 2009.

34 *his first paper:* David W. Keith and Hadi Dowlatabadi, "A Serious Look at Geoengineering," *Eos: Transactions of the American Geophysical Union* 73 (1992): 289–93.

35 *The NAS report:* National Academy of Sciences, *Policy Implications of Greenhouse Warming: Mitigation, Adaptation, and the Science Base* (Washington, DC: National Academies Press, 1992), pp. 433–64.

35 *"modest":* Interview with Hadi Dowlatabadi, April 2009.

42 *"This is an X-29":* For more about the design of the X-29, see Jim Winchester, *The Encyclopedia of Modern Aircraft* (San Diego: Thunder Bay Press, 2005).

43 *"asked you to ride it":* Interview with Stephen Salter, June 2009.

45 *"how we pursue this":* Interview with David Keith, June 2008.

3. God's Machine

47 *Svante Arrhenius:* For a good primer on the history of global warming science, see Spencer R. Weart, *The Discovery of Global Warming,* 2nd ed. (Cambridge, MA: Harvard University Press, 2008). See also Wallace S. Broecker and Robert Kunzig, *Fixing Climate: What Past Climate Changes Reveal About the Current Threat—and How to Counter It* (New York: Hill and Wang, 2008).

48 *"rapidly propagating mankind":* Cited in Broecker and Kunzig, *Fixing Climate,* p. 69.

49 *"pristine cosmos":* D. H. Lawrence, "The Hopi Snake Dance," in *Mornings in Mexico and Other Essays* (Cambridge, Eng.: Cambridge University Press, 2009), p. 279.

49 *"group fornication":* Margaret Jacobs, "Making Savages of Us All: White Women, Pueblo Indians, and the Controversy over Indian Dances in the 1920s," in *American Nations: Encounters in Indian Country, 1850 to the Present,* ed. Fredrick E. Hoxie (New York: Routledge, 2001), p. 179.

49 *"primal environmental catastrophe":* Kerry Emanuel, *What We Know About Climate Change* (Cambridge, MA: MIT Press, 2007), p. 4.

49 *"all was light":* Quoted in Patricia Fara, *Newton: The Making of Genius* (New York: Columbia University Press, 2002), p. 161.

50 *"God's machine":* Edward O. Wilson, *Consilience: The Unity of Knowledge* (New York: Knopf, 1998), p. 24.

50 *four categories:* Gavin Pretor-Pinney, *The Cloudspotter's Guide: The Science, History, and Culture of Clouds* (New York: Perigee Books, 2006), p. 9.

51 *"indexes of climate":* For the best overview of early Americans' interest in weather, see James Fleming, *Historical Perspectives on Climate Change* (New York: Oxford University Press, 2005). See also William Eleroy Curtis, *The True Thomas Jefferson* (New York: J. B. Lippincott, 1904).

51 *James Pollard Espy:* The details of Espy's career are from James R. Fleming, "The Climate Engineers," *Wilson Quarterly* (Spring 2007): 45–60, as well as Fleming's "The Pathological History of Climate and Weather Modification: Three Cycles of Promise and Hype," *Historical Studies in the Physical and Biological Sciences* 37, no. 1 (2006): 4. For an excellent overview of the rainmaking era, see Clark C. Spence, *The Rainmakers: American "Pluviculture" to World War II* (Lincoln: University of Nebraska Press, 1980), pp. 9–22.

51 *"steam power":* James R. Fleming, "The Climate Engineers," p. 49.

52 *"Magnificent Humbug":* Ibid., p. 5.

53 *"pluviculturists":* Spence, *The Rainmakers*, p. 2.

53 *Dr. George Ambrosius Immanuel Morrison Sykes:* Much more about Sykes's strange inventions can be found in Raymond Hulbert, "Science Still Seeks a Rain-Making Machine," *Modern Mechanics*, January 1931. See also Spence, *The Rainmakers*, pp. 129–34.

53 *"eats out of my hand":* Quoted in Spence, *The Rainmakers*, p. 131.

54 *"like a sponge":* Quoted ibid., p. 56.

54 *"four hundred percent bigger":* Quoted in Garry Jenkins, *The Wizard of Sun City: The Strange True Story of Charles Hatfield, the Rainmaker Who Drowned a City's Dreams* (New York: Thunder's Mouth Press, 2005), p. 51.

54 *"the most inventive decade":* Vaclav Smil, *Transforming the Twentieth Century: Technical Innovations and Their Consequences* (New York: Oxford University Press, 2006), p. 14.

54 *"two fishes":* Jonathan Raban, *Bad Land: An American Romance* (New York: Vintage, 1997), p. 153.

55 *"farm or garden":* Quoted in "The Geography of Hope," *The West*, Ken Burns, executive producer; Stephen Ives, director (PBS, 1996).

55 *"the earth and the air":* Spence, *The Rainmakers*, p. 101.

55 *"psychology of drought":* Quoted ibid., p. 8.

56 *Charles Hatfield:* Jenkins, *The Wizard of Sun City*, pp. 17–30.

56 *"Henry Ford of rainmaking":* Quoted ibid., p. 6.

57 *"atmosphere of the universe":* Ibid., pp. 6–40.

57 *"healthful atmosphere":* Quoted ibid., p. 19.

59 *"God being within":* Ibid., p. 39.

59 *"He was a dandy":* Quoted ibid., p. 17.

59 *"science of the atmosphere":* Quoted ibid., p. 36.

60 *"Fume men":* Spence, *The Rainmakers*, p. 81.

60 *"a moisture accelerator":* Quoted ibid.

60 *"mustard plaster":* Quoted in Jenkins, *The Wizard of Sun City,* p. 90.

60 *"Limburger cheese factory":* Quoted ibid.

61 *"Frankenstein of the air":* Quoted ibid., p. 48.

61 *"by natural means":* Quoted ibid., p. 43.

63 *"make some rainfall":* Quoted ibid., p. 71.

63 *"I ask no compensation":* Quoted ibid., p. 72.

64 *"Is the Rainmaker at Work?":* Quoted ibid., p. 131.

65 *"awaiting the laurel wreath":* Quoted ibid., p. 207.

66 *high-profile scientific paper:* Kerry Emanuel, "Increasing Destructiveness of Tropical Cyclones over the Past Thirty Years," *Nature* 436 (July 31, 2005): 686–88.

67 *"tomatoes don't ripen":* Interview with Richard Alley, November 2006.

68 *scathing essay:* David Starr Jordan, "The Art of Pluviculture," *Science* 62 (July 25, 1925): 81–82.

4. Big Science

70 *Nevada Test Site:* For more information, see "Nevada Test Site," U.S. Department of Energy, http://www.nv.doe.gov/nts/default.htm.

71 *"common sense":* Steven Shapin, "Megaton Man," *New York Review of Books,* April 25, 2002.

71 *"change the earth's surface":* Quoted in Scott Kirsch, *Proving Grounds: Project Plowshare and the Unrealized Dream of Nuclear Earthmoving* (New Brunswick, NJ: Rutgers University Press, 2005), p. 3.

72 *"incandescent gases":* Dan O'Neill, *The Firecracker Boys: H-Bombs, Inupiat Eskimos, and the Roots of the Environmental Movement* (New York: Basic Books, 2007), p. 274.

72 *"earth wide open":* Interview with Frank Sullivan, March 2009.

73 *"explosive reality":* Shapin, "Megaton Man."

76 *"razing mountains":* Quoted in O'Neill, *The Firecracker Boys,* p. 24.

76 *"alter its climate":* N. Rusin and L. Flit, *Man Versus Climate* (Moscow: Peace Publishers, 1960), p. 17.

76 *"enormous areas":* Ibid., p. 54.

77 *"bombs into plowshares":* Quoted in O'Neill, *The Firecracker Boys,* p. 27.

77 *"work with reality":* Quoted in Kirsch, *Proving Grounds,* p. 65.

77 *"in future generations":* Quoted ibid., p. 33.

77 *"confused manner":* Edward Teller, *The Legacy of Hiroshima* (New York: Doubleday, 1962), p. 174.

77 *atmospheric testing:* These statistics are cited in O'Neill, *The Firecracker Boys,* p. 33.

77 *"geographical engineering":* Teller, *The Legacy of Hiroshima,* p. 84.

78 *"blast the ice pack":* Ibid., p. 88.

78 *"this unique substance":* Ibid., p. 89.

79 *"necessary precautions":* Ibid., p. 85.

79 *"requires courage":* Teller, *The Legacy of Hiroshima,* p. 79.

81 *"first victims"*: Quoted in O'Neill, *The Firecracker Boys*, p. 97.

81 *"all mankind"*: Quoted in Kirsch, *Proving Grounds*, p. 54.

82 *"coal miners"*: Quoted in O'Neill, *The Firecracker Boys*, p. 37.

82 *"obvious demonstration"*: Quoted in Kirsch, *Proving Grounds*, p. 76.

82 *"human geographical studies"*: Quoted ibid., p. 80.

83 *"totally abandoned"*: Quoted ibid., p. 97.

83 *"alter the earth"*: Paul Brooks and Joseph Foote, "The Disturbing Story of Project Chariot," *Harper's*, April 19, 1962, p. 60.

84 *"95 percent"*: Quoted in Kirsch, *Proving Grounds*, p. 133.

84 *"5 times as much"*: Quoted in O'Neill, *The Firecracker Boys*, p. 275.

85 *"biological insanity"*: Quoted in Kirsch, *Proving Grounds*, p. 178.

85 *"termination of Plowshare"*: Quoted in O'Neill, *The Firecracker Boys*, p. 295.

85 *"all things Russian"*: Shapin, "Megaton Man."

86 *"human frailties"*: Lowell Wood, remarks, *Newsline* (employee newsletter of Lawrence Livermore National Laboratory), September 12, 2003.

86 *"change the earth's surface"*: Quoted in Kirsch, *Proving Grounds*, p. 3.

87 *"with no purpose"*: Interview with Lowell Wood, September 2006.

87 *"If the sea levels"*: Quoted in Collin Levy, "Rocket Man," *Wall Street Journal*, January 5, 2008.

5. The Blue Marble

89 *"his ideas were heresy"*: Interview with Ken Caldeira, June 2007.

89 *"about to fail"*: James Lovelock, *The Revenge of Gaia* (London: Allen Lane, 2006), p. 6.

91 *"cunning mixture"*: "Gaia Worship—the New Pagan Religion," post, Contender Ministries, http://contenderministries.org/UN/gaia.php.

91 *"eliminate us"*: James Lovelock, *Homage to Gaia: The Life of an Independent Scientist* (Oxford: Oxford University Press, 2000), p. 376.

93 *London neighborhood*: Lovelock recounts his childhood in *Homage to Gaia*, pp. 7–37. See also John Gribbin and Mary Gribbin, *James Lovelock: In Search of Gaia* (Princeton, NJ: Princeton University Press, 2009), pp. 20–47.

94 *"War is evil"*: Lovelock, *Homage to Gaia*, p. 50.

95 *"no conceivable hazard"*: Ibid., p. 216.

96 *"poisonous sun"*: Interview with Paul Ehrlich, January 2007.

98 *"cybernetics"*: The classic account of cybernetic theory is Norbert Wiener's *Cybernetics: Or Control and Communication in the Animal and the Machine* (Cambridge, MA: MIT Press, 1965).

98 *Why not Gaia?*: Lovelock's account of his time at JPL and his inspiration for Gaia can be found in *Homage to Gaia*, pp. 241–55. See also Gribbin and Gribbin, *James Lovelock*, pp. 137–62.

98 *"inner circle of the sun"*: James Lovelock, *Gaia: A New Look at Life on Earth* (New York: Oxford University Press, 2000), p. 11.

98 *"pop ecology literature":* Richard Dawkins, *The Extended Phenotype: The Long Reach of the Gene* (Oxford: Oxford University Press, 1982), p. 162.

98 *"an evil religion":* Quoted in James Lovelock, *The Vanishing Face of Gaia: A Final Warning* (New York: Basic Books, 2009), p. 174.

100 *"dynamic nature of the planet":* Interview with Wally Broecker, January 2008.

100 *"single, self-regulating system":* Earth System Science Partnership, "The Amsterdam Declaration on Global Change" (Global Change Open Science Conference, Amsterdam, July 13, 2001), http://www.essp.org/index.php?id=41. For more about Earth System Science, see Lee R. Kump, James F. Kasting, and Robert G. Crane, *The Earth System* (Upper Saddle River, NJ: Pearson, 2004).

101 *"the world as such":* Václav Havel, "The Need for Transcendence in the Modern World" (speech, Independence Hall, Philadelphia, July 4, 1994).

103 *three feet or more:* Anil Ananthaswamy, "Sea Level Rise: It's Worse Than We Thought," *New Scientist,* July 1, 2009.

104 *hotter stable state:* Lovelock's idea that the earth's climate could jump to a hotter stable state remains controversial among climate scientists. As Ken Caldeira wrote in a private online discussion in November 2009 (quoted here with permission),

> "This idea of two stable states has been pushed recently by Jim Lovelock. However I think there is very little evidence to support such a view. Just about every model result indicates monotonically and more-or-less continuously increasing effect with increasing doses of CO_2 (or changes in sunlight). [Of course, the models are a gross simplification of reality.] There are a few sources of metastability (e.g., large ice sheets) but I do not think that these sources of metastability govern the overall behavior of the system on sub-millennial time scales."

104 *other complex systems:* Marten Scheffer et al., "Early Warning Signals for Critical Transitions," *Nature* 461 (September 3, 2009): 53–59.

106 *"use the small input":* James Lovelock, "Nuclear Power Is the Only Green Solution," *Independent,* May 24, 2004.

106 *"earth has a fever":* Al Gore, interview, *Good Morning America,* June 23, 2006.

106 *"Russian roulette":* Personal communication, June 2007.

107 *"political emergency":* Interview with David Victor, August 2009.

6. Doping the Stratosphere

109 *anything but inspiring:* The full scientific program for the 2008 AGU meeting, including a videotape of James Hansen's lecture, can be found at http://www.agu .org/webcast/fm08/.

110 *devoid of life:* Some scientists believe that Hansen is overstating the case. When I queried Ray Pierrehumbert, at the University of Chicago, who has done a lot of research on the atmospheres of other planets, about Hansen's claims, he emailed me this response: "It's true that you can't completely rule out a runaway [warm-

ing] based on a priori physics bounds, but for it to happen you need all the low clouds to disappear and the atmosphere to become saturated with moisture. Very far-fetched, and if this didn't happen during earlier warm periods like the PETM [Paleocene-Eocene Thermal Maximum] it's really implausible that it would happen as a result of anthropogenic warming."

111 *"It's wrong to mug":* Interview with Ken Caldeira, December 2008.

112 Rolling Stone: Jeff Goodell, "Can Dr. Evil Save the World?" *Rolling Stone*, November 3, 2006.

112 *"could be engineered":* Malcolm Gladwell, "In the Air," *The New Yorker*, May 12, 2008.

113 *"reasonably crazy people":* Interview with Bill Gates, October 2009.

116 *dreaming about such techniques:* For a good overview of military involvement in climate and weather manipulation schemes, see James R. Fleming, "Fixing the Weather and Climate: Military and Civilian Schemes for Cloud Seeding and Climate Engineering," in *The Technological Fix: How People Use Technology to Create and Solve Problems*, ed. Lisa Rosner, pp. 175–200 (New York: Routledge, 2004).

118 *"actually implement them":* Interview with Freeman Dyson, October 2006.

119 *"O Group":* This account of Wood's time at Livermore is drawn from William J. Broad, *Star Warriors: A Penetrating Look into the Lives of Young Scientists Behind Our Space Age Weaponry* (New York: Touchstone, 1986), p. 00.

119 *KGB agents and terrorists:* Interview with Richard Gabriel, October 2006.

121 *his first scientific paper:* Ken Caldeira, "Evolutionary Pressures on Planktonic Production of Atmospheric Sulfur," *Nature* 337 (1989): 632–34.

121 *the first model:* Mikhail Budyko, *The Heat Energy Balance of the Earth's Surface* (Leningrad: Hydrometeorological Publishing House, 1956). English translations of Budyko's text are extremely difficult to find.

121 *mimic volcanoes:* Cited in Hubert H. Lamb, *Climatic History and the Future* (Princeton, NJ: Princeton University Press, 1977), pp. 46, 660–61.

122 *"ocean acidification":* Ken Caldeira and Michael E. Wickett, "Anthropogenic Carbon and Ocean pH," *Nature* 425 (September 25, 2003): 365.

122 *"one of the surprises":* National Academy of Sciences, *Policy Implications of Greenhouse Warming*, p. 460.

123 *"I was stunned":* Interview with David Keith, January 2008.

124 *Caldeira's paper:* Bala Govindasamy and Ken Caldeira, "Geoengineering Earth's Radiation Balance to Mitigate CO_2-Induced Climate Change," *Geophysical Research Letters* 27 (2000): 2141–44.

124 *"counteracting climate change":* Ibid., p. 2144.

125 *"People never pay":* Email correspondence with Lowell Wood, October 23, 2006.

126 *"on Our Side!":* Email correspondence with Lowell Wood, October 26, 2006. The United Nations Framework Convention on Climate Change (UNFCC), called the Rio Framework Convention for short, is the international environmental treaty that resulted from the Earth Summit held in Rio de Janeiro in 1992. The objective of the treaty was to stabilize greenhouse gas concentrations in the at-

mosphere at a level that would prevent dangerous anthropogenic interference with the climate system. The Kyoto Protocol, which was adopted in 1997, is essentially a refinement and update of the UNFCC.

126 *"doing less is catastrophic"*: Interview with Gregory Benford, September 2006.

128 *finished their model runs*: Wood and Caldeira's collaboration eventually resulted in a paper: Ken Caldeira and Lowell Wood, "Global Arctic Climate Engineering: Numerical Model Studies," *Philosophical Transactions of the Royal Society* 366 (November 2008): 4039–56.

129 *In one recent study*: J. J. Blackstock et al., "Climate Engineering Responses to Climate Emergencies" (Novim, 2007), p. 46, http://arvix.org/pdf/0907.5140.

129 *recently floated*: Personal communication, July 2009.

131 *"exceedingly complex"*: Interview with Michael Oppenheimer, October 2006.

131 *impact on ozone levels*: Simone Tilmes et al., "The Sensitivity of Polar Ozone Depletion to Proposed Geo-Engineering Schemes," *Science* 320 (May 30, 2008): 1201–4.

131 *eruption of Mount Pinatubo*: Paul Telford et al., "Reassessment of Causes of Ozone Column Variability Following the Eruption of Mount Pinatubo Using a Nudged CCM," *Atmospheric Chemistry and Physics* 9 (2009): 4251–60. See also Neil Harris et al., "Trends in Stratospheric and Tropospheric Ozone," *Journal of Geophysical Research* 102 (1997): 1571–90.

131 *300 million years*: Caldeira and Wickett, "Anthropogenic Carbon and Ocean pH," p. 365.

131 *"just beginning to understand"*: Interview with Stephen Schneider, August 2008.

133 *"take the edge off"*: T.M.L. Wigley, "A Combined Mitigation/Geoengineering Approach to Climate Stabilization," *Science* 314 (October 20, 2006): 452–54.

134 *"Comrades"*: Email correspondence with Wood, October 26, 2006.

7. A Little Cash on the Side

135 *"machinery works"*: Quoted in Graham Lawton, "Interview: America Turns Red, White and Green," *New Scientist*, August 3, 2009.

136 *photosynthetic life*: Sallie W. Chisholm et al., "Dis-crediting Ocean Fertilization," *Science* 294 (October 12, 2001): 309.

137 *the biological pump*: J. A. Raven and P. G. Falkowski, "Ocean Sinks for Atmospheric CO_2," *Plant, Cell and Environment* 22, no. 6 (1999): 741–55.

137 *on the* Knorr: You can learn more about the *Knorr*, as well as track the ship's movements in real time, at the Woods Hole website, http://www.whoi.edu/page .do?pid=8157.

139 *Martin eventually hypothesized*: John H. Martin and Steve E. Fitzwater, "Iron Deficiency Limits Phytoplankton Growth in the North-east Pacific," *Nature* 331 (January 28, 1988): 341–43. For a general summary of Martin's life and ideas, see "John Martin," NASA Earth Observatory, http://earthobservatory.nasa.gov/ Features/Martin/martin_3.php.

139 *"I'll give you an ice age"*: Quoted in Hugh Powell, "Will Ocean Fertilization Work?" *Oceanus*, January 2008, p. 37.

140 *$3 trillion by 2020:* "Carbon Market Transactions in 2020: Dominated by Financials?" (report, Point Carbon, May 22, 2008). It's worth noting that the report assumes a very optimistic scenario, with thirty-eight billion tons of CO_2 equivalent being swapped in a global market at a price of about $78 per ton.

140 *"brain damage"*: David Brower, "It's Healing Time on Earth" (Twelfth Annual E. F. Schumacher Lecture, Stockbridge, MA, October 1992.

141 *cut sulfur dioxide emissions:* U.S. Environmental Protection Agency, "Acid Rain and Related Programs: 2008 Highlights" (December 2009), p. 3.

141 *"green success story"*: "The Invisible Green Hand," *Economist*, July 4, 2002.

143 *a hundred times less:* Quirin Schiermeier, "The Oresmen," *Nature* 421 (January 9, 2003): 110.

145 *"ecologist wanderer"*: Planktos website, http://www.planktos-science.com/history.html (accessed April 2007; site now discontinued).

145 *"Greenpeace ship Rainbow Warrior"*: Russ George, "I Am Not the Enemy," *Ottawa Citizen*, June 22, 2007.

146 *"true economically and socially"*: Quoted in Connie Hargrave, "Cold Fusion: A Glimpse into the Future," *Share International*, November 1992.

146 *HaidaClimate:* George, "I Am Not the Enemy."

146 *"the carbon trading world"*: Quoted in "Partner Ecosystem Restoration Companies Achieve New Project Milestones Enabling Large Volume, Low Cost Carbon Credit Sales," *Business Wire*, April 30, 2007.

147 *wooden schooner Ragland:* Kalee Thompson, "Carbon Discredit," *Popular Science*, July 2008. You can view a short movie of Russ George's trip on the *Ragland* at http://www.planktos-science.com/history.html.

147 *Ocean Technology Group:* Chisholm et al., "Dis-crediting Ocean Fertilization."

147 *offered him $1 million:* Carrie Peyton Dahlberg, "SF Entrepreneur Floats Bold Idea to 'Fertilize' Ocean," *Sacramento Bee*, March 30, 2008.

148 *escaped going to jail:* Peter Kennedy, "Skalbania Edges Way Back from Sidelines," *Globe and Mail* (Toronto), November 17, 2003.

148 *"getting it financed"*: Quoted ibid.

148 *"good for the world"*: Interview with Nelson Skalbania, February 2009.

148 *$750 million:* Matt Richtel, "The Plankton Defense," *New York Times*, May 1, 2007.

148 *top ten bachelors:* Janelle Brown, "Valley of the Dolls," *Salon*, January 20, 2000, http://www.salon.com/tech/log/2000/01/20/eligible_bachelors/index.html.

148 *fertilize the oceans:* Interview with Dan Whaley, June 2009.

149 *stealing company secrets:* Interviews with Whaley and Skalbania, 2009.

149 *"The oceans are dying"*: Interview with Russ George, August 2006.

151 *"quite simply, not true"*: Chisholm et al., "Dis-crediting Ocean Fertilization."

152 *between 5 and 50 percent:* Ken O. Buesseler and Phillip W. Boyd, "Will Ocean Fertilization Work?" *Science* 300 (April 4, 2003): 67–68.

152 *more like 4 percent:* Ken Caldeira, "Iron Fertilization: Ocean Science Meets Political Controversy" (PowerPoint presentation, June 2009). In the presentation, Caldeira cited Raymond Pollard et al., "Ocean Fertilization Likely to Be Marginally Effective in Storing CO_2 in the Ocean," *Nature* 457 (January 29, 2009), pp. 577–80.

153 *invested $1 million:* interview with Peter Willcox, February 2009.

153 *"crazy mascot":* Quoted in John Laumer, "Planktos Inc., 'Seeds of Iron' to Mitigate Climate Change," TreeHugger, February 21, 2006, http://www.treehugger.com/files/2006/02/planktos.php.

154 *"truly meaningful hope":* Noel Brown, remarks (press conference for Planktos "Voyage of Discovery," National Press Club, Washington, DC, March 14, 2007), http://www.youtube.com/watch?v=18acL_kKTJs.

154 *"planet-friendly tool":* Planktos website, www.planktos.com (accessed April 2007; site now discontinued).

154 *"Galápagos Seas":* Cited in Thompson, "Carbon Discredit."

155 *"global warming snake oil":* Quoted in Andrew Revkin, "Project to Capture CO_2 with Plankton Puts to Sea," *New York Times*, Dot Earth blog, November 6, 2007, http://dotearth.blogs.nytimes.com/2007/11/06/project-to-harness-plankton-puts-to-sea/.

155 *"engineered nanoparticles":* Quoted in Thompson, "Carbon Discredit."

155 *safe and natural:* Ibid. See also George, "I Am Not the Enemy."

155 *"a dangerous experiment":* Sea Shepherd Conservation Society, "Operation Aquatic Dust Storm" (press release, August 10, 2007).

156 *"false and manipulative":* International Environmental Law Project to Ms. Linda Thompson, Director, Division of Enforcement, Securities and Exchange Commission, September 17, 2007, acquired from anonymous source.

156 *"nature of the research":* Quoted in Thompson, "Carbon Discredit."

156 *Andrew Revkin:* Revkin, "Project to Capture CO_2."

157 *French secret service:* Interview with Peter Willcox, February 2009. See also Michael Small, "His First Ship Was Bombed by the French, but Peter Willcox Won't Give Up the Fight," *People*, March 26, 1990.

158 *"a perfect storm":* Interview with Michael Bailey, February 2009.

158 *"disinformation campaign":* Quoted in David Baines, "Environmental Crusaders Sink Instead of Swim," *Vancouver Sun*, March 15, 2008.

159 *Planktos meltdown:* Interview with Dan Whaley, April 2008.

161 *"We're still interested":* Interview with Dan Whaley, June 2009.

161 *"to stimulate and support":* Climate Response Fund website, http://climateresponsefund.org/.

161 *"The whole idea":* email from David Keith, March 2009.

162 *"If we decide":* Email from David Keith, January 2010.

162 *"tremendous number of antibodies":* Interview with Lowell Wood, December 2008.

162 *"a headache":* Interview with Ken Caldeira, June 2009.

8. The Romance of Clouds

164 *Cloud Appreciation Society:* For more about the society, go to http:// cloudappreciationsociety.org/.

164 *"Nothing in nature":* Pretor-Pinney, *The Cloudspotter's Guide*, p. 9.

165 *"the atmosphere's moods":* "The Manifesto of the Cloud Appreciation Society," http://cloudappreciationsociety.org/manifesto/.

166 *"greasy-fingered mechanic":* Interview with Stephen Salter, June 2009.

168 *image of the vessel:* Several pictures of Salter's ship, as well as earlier examples of ships equipped with Flettner rotors, can be see in Stephen Salter and John Latham, "Sea-Going Hardware for the Cloud Albedo Method of Reversing Global Warming," *Philosophical Transactions of the Royal Society* 366, no. 1882 (November 13, 2008): 3989–4006.

170 *four basic types:* Pretor-Pinney, *The Cloudspotter's Guide*, p. 52.

170 *"observation and naming":* Quoted in Richard Holmes, *The Age of Wonder: How the Romantic Generation Discovered the Beauty and Terror of Science* (New York: Pantheon Books, 2008), p. 160.

171 *"bombing accuracy":* Stewart Halsey Ross, *Strategic Bombing by the United States in World War II* (New York: McFarland, 2002), p. 104.

171 *encourage ice formation:* Langmuir and Schaefer's work at GE is recounted in Pretor-Pinney, *The Cloudspotter's Guide*, pp. 257–61.

172 *"I shouted":* Quoted in William Langewiesche, "Stealing Weather," *Vanity Fair*, May 2008.

172 *"snow for winter resorts":* "Three Mile Cloud Made into Snow by Dry Ice Dropped from Plane," *New York Times*, November 15, 1946, p. 24.

172 *"Until the day he died":* Langewiesche, "Stealing Weather."

173 *"a hundred million volts":* Interview with John Latham, July 2009.

175 *Slingo argued:* Anthony Slingo, "Sensitivity of the Earth's Radiation Budget to Changes in Low Clouds," *Nature* 343 (January 4, 1990): 49–51.

176 *reflect more sunlight:* Sean Twomey, "The Influence of Pollution on the Shortwave Albedo of Clouds," *Journal of Atmospheric Sciences* 34 (1977): 1149–52.

176 *"the net effect":* Spencer Weart, *The Discovery of Global Warming*. The chapter concerning aerosols is available at http://www.aip.org/history/climate/aerosol.htm.

176 *"The essence":* Interview with Alan Gadian, July 2009.

177 *Latham's idea:* John Latham, "Control of Global Warming?" *Nature* 347 (September 27, 1990): 349–50.

178 *"deployment of these rainmakers":* Quoted in Rob Edwards, "Floating Wind Turbines Could Whip Up Rain," *New Scientist*, May 25, 2002.

182 *nearly half the warming:* Andy Jones et al., "Climate Impacts of Engineering Marine Stratocumulus Clouds," *Journal of Geophysical Research* 114 (May 27, 2009): 000–00.

183 *"air-condition Abu Dhabi":* Interview with Ken Caldeira, July 2009.

185 *sharply decrease rainfall:* Jones et al., "Climate Impacts of Engineering Marine Stratocumulus Clouds."

185 *no evidence of serious effects:* John Latham, Phil Rasch, et al., "Global Temperature Stabilization via Controlled Albedo Enhancement of Low-Level Maritime
Clouds," *Philosophical Transactions of the Royal Society* 366 (2008): 3969–87.

185 *"we can perturb clouds":* Interview with Phil Rasch, December 2008.

186 *"The main uncertainty":* Email correspondence with Ray Pierrehumbert, October 2009. It's important to note that Pierrehumbert, an articulate and forceful
critic of many geoengineering schemes, believes that learning how to brighten
clouds is less important than learning how clouds work. "Albedo engineering
info should be the spinoff of something you would do anyway for important
cloud problems, not the other way around," he wrote.

186 *$20 million:* Interview with Kelly Wanser, September 2009.

187 *blasted the report:* ETC Group, "The Emperor's New Climate: Geoengineering as
21st Century Fairytale" (special report, August 28, 2009).

188 *"acting in bad faith":* Ken Caldeira, post, Google discussion group on geoengineering, August 29, 2009, http://groups.google.com/group/geoengineering?hl=en.

188 *"All it takes":* Interview with David Victor, August 2009.

189 *"This is an argument":* Email correspondence with Stephen Salter, August 16,
2009.

9. A Global Thermostat

190 *two-day conference:* Climate Engineering Workshop, Harvard University, November 8–9, 2007.

191 *"white men":* Interview with Sallie Chisholm, November 2007.

191 *"He had no idea":* Others had a different view of Summers's comments at the
Harvard conference. David Keith later told me via email (November 2009),
"Summers's main argument was that it's arrogant for the science community to
think that they can keep [geoengineering] a secret because they are afraid people will lose their focus on mitigation. He spoke forcibly and brilliantly on this
point."

195 *"Unilateral deployment":* Interview with David Victor, August 2009.

196 *one thousand billionaires:* "The World's Billionaires," *Forbes,* March 11, 2009,
http://www.forbes.com/2009/03/11/worlds-richest-people-billionaires-2009
-billionaires_land.html.

196 *"Greenfingers":* David Victor, "On the Regulation of Geoengineering," *Oxford Review of Economic Policy* 24, no. 2 (2008): 322–36.

198 *"quantum leaps":* Quoted in Jennifer Hattam, "We Are Plenty Good Enough,"
Sierra, November–December 2003, http://www.sierraclub.org/sierra/200311/
interview.asp.

200 *"Troublesome and high-risk":* Kevin Kelly, "Dealing with Rogue Technologies," *The
Technium,* February 24, 2007, http://www.kk.org/thetechnium/archives/2007/02/
dealing_with_ro.php.

202 *"a tailored climate":* Interview with Ken Caldeira, July 2009.

203 *Large Hadron Collider:* Scott Barrett, "The Incredible Economics of Geoengineering," *Environmental Resource Economics* 38 (2008): 51.

203 *"hyperdense sphere":* Quoted ibid.

203 *"really want to test them":* Ibid.

204 *"Atlanta and Moscow":* Ibid., p. 52.

205 *"an especially fortunate outcome":* Ibid.

205 *"something like parity":* Garret Keizer, "Climate, Class, and Claptrap," *Harper's*, June 2007, p. 10.

206 *"in an ice storm":* Personal communication, December 2008.

206 *"suffered from bad weather":* Interview with Bill Gates, October 2009.

207 *"Chemtrails provide":* Douglas Herman, "Chemtrails 101—an Introduction," Rense.com, September 6, 2008, http://www.rense.com/general83/chemm.htm.

207 *"Just because you're paranoid":* James Atlas (Delmore Schwartz's biographer), personal communication, 2002. Another version of this quote: "Even paranoids have real enemies."

208 *"jiggling the planet":* Quoted in Norman Macrae, *John von Neumann* (New York: Random House, 1993), p. 11.

209 *"they will be exploited":* John von Neumann, "Can We Survive Technology?" *Fortune*, June 1955, p. 46.

209 *"a macro-pathological turn":* James R. Fleming, "The Pathological History of Weather and Climate Modification," p. 13.

209 *"weapon of war":* Quoted ibid., p. 14.

209 *Fidel Castro:* Graeme Wood, "Riders on the Storm," *Atlantic*, October 2007.

209 *"a total failure":* Interview with Lowell Wood, December 2008.

209 *issued a report:* Col. Tamzy J. House et al., "The Weather as a Force Multiplier: Owning the Weather in 2025" (August 1996), in *Air Force 2025*, vol. 3, http://csat.au.af.mil/2025/volume3/vol3ch15.pdf.

210 *"war-fighting applications":* Ibid., p. 6.

210 *"natural weather phenomena":* Ibid., p. 36.

210 *"Given such mindsets":* Fleming, "The Climate Engineers," p. 60.

210 *"Recent war games":* John M. Broder, "Climate Change Seen as Threat to National Security," *New York Times*, August 8, 2009.

210 *"It gets real complicated":* Quoted ibid.

211 *"a threat multiplier":* Quoted in Robert McClure, "At SEJ, Gloom and Doom Without a Sense of Humor," *Grist*, October 9, 2009, http://www.grist.org/article/2009-10-09-at-sej-doom-and-gloom-without-the-sense-of-humor/.

211 *"The military can muster":* Interview with Gregory Benford, August 2009.

213 *"NASA is far better":* Interview with David Keith, October 2009. The NASA program he is referring to was known as High Speed Research and was phased out in the late 1990s.

214 *"drink to getting hard":* Andrei Sakharov, *Memoirs* (New York: Knopf, 1990), p. 194.

10. Human Nature

216 *"Only God"*: Don Blankenship, speech (Friends of America rally, Charleston, WV, September 7, 2009), http://www.youtube.com/watch?v=MdU1PV8qAvY.

216 *"All hurricanes"*: John Hagee, *Fresh Air*, National Public Radio, September 18, 2006.

217 *"approach very cautiously"*: Interview with Mike MacCracken, June 2009.

218 *"a nasty bitch"*: Lee M. Silver, *Challenging Nature: The Clash of Science and Spirituality at the New Frontiers of Life* (New York: HarperCollins, 2006), p. 202.

219 *"miracles become"*: Charles Darwin, *The Autobiography of Charles Darwin* (New York: W. W. Norton, 1993), p. 86.

220 *"its own stupidity"*: Rowan Williams, "Renewing the Face of the Earth: Human Responsibility and the Environment" (address, York Minster, March 25, 2009).

221 *"the emptiest deal"*: "Dismal Outcome at Copenhagen Fiasco," *Financial Times*, December 20, 2009.

221 *"a humanitarian crisis"*: Quoted in Louise Gray, "The Copenhagen Climate Conference: Who Is Going to Save the Planet Now?" *Daily Telegraph* (London), December 21, 2009.

222 *Klaus Lackner*: For a good account of Lackner's efforts with air capture, see Broecker and Kunzig, *Fixing Climate*, pp. 197–232.

222 *Peter Eisenberger*: Nicola Jones, "Climate Crunch: Sucking It Up," *Nature* 458 (April 30, 2009): 1094–97.

223 *emailed me a list*: Email correspondence with Ken Caldeira, November 2009.

225 *"Copenhagen wouldn't be necessary"*: Quoted in Andrew Revkin, "Branson on the Power of Biofuels and Elders," *New York Times*, Dot Earth blog, October 15, 2009, http://dotearth.blogs.nytimes.com/2009/10/15/branson-on-space-climate-biofuel-elders/.

225 *"Thinking seriously"*: Interview with Dan Dudek, November 2009.

226 *one late-night conversation*: Interview with Ken Caldeira, September 2009.

226 *"less cleverness"*: Quoted in Kai Bird and Martin J. Sherwin, *American Prometheus: The Triumph and Tragedy of J. Robert Oppenheimer* (New York: Knopf, 2005), p. 560.

Selected Bibliography

Archer, David. *The Long Thaw: How Humans Are Changing the Next 100,000 Years of Earth's Climate.* Princeton, NJ: Princeton University Press, 2009.

Arthur, W. Brian. *The Nature of Technology: What It Is and How It Evolves.* New York: Free Press, 2009.

Bird, Kai, and Martin J. Sherwin. *American Prometheus: The Triumph and Tragedy of J. Robert Oppenheimer.* New York: Knopf, 2005.

Brand, Stewart. *Whole Earth Discipline: An Ecopragmatist Manifesto.* New York: Viking, 2009.

Broad, William J. *Star Warriors: A Penetrating Look into the Lives of Young Scientists Behind Our Space Age Weaponry.* New York: Touchstone, 1986.

Broecker, Wallace S., and Robert Kunzig. *Fixing Climate: What Past Climate Changes Reveal About the Current Threat—and How to Counter It.* New York: Hill and Wang, 2008.

Budyko, Mikhail. *The Heat Balance of the Earth's Surface.* Leningrad: Hydrometeorological Publishing House, 1956.

Dawkins, Richard. *The Extended Phenotype: The Long Reach of the Gene.* Oxford: Oxford University Press, 1982.

Donnan, Jack, and Marcia Donnan. *Rain Dance to Research: The Story of Weather Control.* New York: David McKay, 1977.

Dyson, Freeman. *Disturbing the Universe.* New York: Basic Books, 1979.

Emanuel, Kerry. *What We Know About Climate Change.* Cambridge, MA: MIT Press, 2007.

Flannery, Tim. *The Weather Makers: How Man Is Changing the Climate and What It Means for Life on Earth.* New York: Atlantic Monthly Press, 2005.

Fleming, James Rodger. *Historical Perspectives on Climate Change.* New York: Oxford University Press, 2005.

Goodell, Jeff. *Big Coal: The Dirty Secret Behind America's Energy Future.* New York: Houghton Mifflin, 2006.

Gribbin, John, and Mary Gribbin. *James Lovelock: In Search of Gaia.* Princeton, NJ: Princeton University Press, 2009.

Holmes, Richard. *The Age of Wonder: How the Romantic Generation Discovered the Beauty and Terror of Science.* New York: Pantheon Books, 2008.

Jenkins, Garry. *The Wizard of Sun City: The Strange True Story of Charles Hatfield, the Rainmaker Who Drowned a City's Dreams.* New York: Thunder's Mouth Press, 2005.

Kelly, Cynthia C., ed. *The Manhattan Project: The Birth of the Atom Bomb in the Words of Its Creators, Eyewitnesses, and Historians.* New York: Black Dog and Leventhal, 2007.

Kelly, Kevin. *Out of Control: The New Biology of Machines, Social Systems, and the Economic World.* New York: Addison-Wesley, 1994.

Kirsch, Scott. *Proving Grounds: Project Plowshare and the Unrealized Dream of Nuclear Earthmoving.* New Brunswick, NJ: Rutgers University Press, 2005.

Kolbert, Elizabeth. *Field Notes from a Catastrophe: Man, Nature, and Climate Change.* New York: Bloomsbury, 2006.

Kump, Lee R., James F. Kasting, and Robert G. Crane. *The Earth System.* Upper Saddle River, N.J.: Pearson, 2004.

Lawrence, D. H. *Mornings in Mexico and Other Essays.* Cambridge, Eng.: Cambridge University Press, 2009.

Leach, Edmund. *A Runaway World?* Oxford: Oxford University Press, 1968.

Lovelock, James. *The Ages of Gaia: A Biography of Our Living Earth.* 2nd ed. Oxford: Oxford University Press, 2000.

———. *Gaia: A New Look at Life on Earth.* Oxford: Oxford University Press, 2000.

———. *Homage to Gaia: The Life of an Independent Scientist.* Oxford: Oxford University Press, 2000.

———. *The Revenge of Gaia: Earth's Climate Crisis and the Fate of Humanity.* New York: Basic Books, 2006.

———. *The Vanishing Face of Gaia: A Final Warning.* New York: Basic Books, 2009.

Macrae, Norman. *John von Neumann.* New York: Random House, 1993.

O'Neill, Dan. *The Firecracker Boys: H-Bombs, Inupiat Eskimos, and the Roots of the Environmental Movement.* New York: Basic Books, 2007.

Pretor-Pinney, Gavin. *The Cloudspotter's Guide: The Science, History, and Culture of Clouds.* New York: Perigee Books, 2006.

Raban, Jonathan. *Bad Land: An American Romance.* New York: Vintage, 1997.

Rosner, Lisa, ed. *The Technological Fix: How People Use Technology to Create and Solve Problems.* New York: Routledge, 2004.

Rowley, Graham. *Cold Comfort: My Love Affair with the Arctic.* Montreal: McGill–Queen's University Press, 2007.

Rusin, Nikolai, and Liya Flit. *Man Versus Climate.* Moscow: Peace Publishers, 1962.

Sakharov, Andrei. *Memoirs.* New York: Knopf, 1990.

Schelling, Thomas C. *Strategies of Commitment and Other Essays.* Cambridge, MA: Harvard University Press, 2006.

Silver, Lee M. *Challenging Nature: The Clash of Science and Spirituality at the New Frontiers of Life.* New York: HarperCollins, 2006.

Smil, Vaclav. *Transforming the Twentieth Century: Technical Innovations and Their Consequences.* New York: Oxford University Press, 2006.

Spence, Clark C. *The Rainmakers: American "Pluviculture" to World War II.* Lincoln: University of Nebraska Press, 1980.

Teller, Edward. *The Legacy of Hiroshima.* New York: Doubleday, 1962.

Verne, Jules. *The Purchase of the North Pole.* 1890. Reprint, New York: Ace Books, 1959.

Weart, Spencer R. *The Discovery of Global Warming.* 2nd. ed. Cambridge, MA: Harvard University Press, 2008.

Wilson, Edward O. *Consilience: The Unity of Knowledge.* New York: Knopf, 1998.

Winchester, Jim. *The Encyclopedia of Modern Aircraft.* San Diego: Thunder Bay Press, 2006.

Wright, Ronald. *A Short History of Progress.* New York: Carroll & Graf, 2004.

Index